Transport & Fate of Chemicals in Soils Principles & Applications

Transport & Fate of Chemicals in Soils Principles & Applications

Editor

Prahlad Maurya

scitus
academics

Transport & Fate of Chemicals in Soils Principles & Applications

Edited by **Prahlad Maurya**

Printed in 2017

ISBN: 978-1-68117-019-0
Library of Congress Control Number: 2015931388

SCITUS Academics LLC,
616, Corporate Way, Suite 2, 4766,
Valley Cottage, NY 10989

www.scitusacademics.com

Contents

Preface

The weathering of the parent material by water determines, to a large extent, the chemical composition of the soil which has ultimately been produced. Some chemicals are leached into the lower soil layers where they accumulate. Other chemicals, more insoluble, are left in the upper layers of the soil. The most rapid removed chemicals are chlorides and sulphates, followed by calcium, sodium, magnesium and potassium. The silicates and oxides of iron and aluminium decompose very slowly and are rarely leached. When some of these products come into contact with the air in the soil, chemical reactions occur, such as oxidation in particular, which results in the formation of chemicals either more soluble or more fragile than the original ones. This results in an acceleration of the weathering processes, increased leaching* of chemicals, and further changes in the soil chemical composition. A freshwater acid sulphate soil may develop when the waterlogged soil containing iron sulphurs (pyrites) is exposed to the air, for example, when constructing ponds. Oxidation of the pyrites takes place and acidification of the soil occurs. The pond water may then become too acid for fish culture. The air present in the soil also contains carbon dioxide. This gas combined with water can form a weak acid (carbonic acid) which will then react with some of the soil chemicals to form new ones. This book, Transport & Fate of Chemicals in Soils: Principles & Applications, offers a comprehensive treatment of the subject complete with supporting examples of mathematical models that describe contaminants reactivity and transport in soils and aquifers. This book details the basic framework of the physical and chemical processes governing the transport of contaminants, trace elements, and heavy metals in soils. Highly practical, it includes laboratory methods, examples, case study, and empirical formulations.

Editor

Chapter 1

Research for Investigating and Managing Soil Contamination Caused by Winter Maintenance in Cold Regions

Helen K. French[1] and Sjoerd E.A.T.M. van der Zee[1]

The Norwegian University of Life Sciences, Department of Plant and Environm. Sciences, NorwayWageningen University, Environmental Sciences Group, Wageningen, The Netherlands

INTRODUCTION

In the north temperate and arctic zones, large amounts of de-icing chemicals are used during the frozen season for winter maintenance of highways, roads, airports and other surface areas. The transport sector, and civil aviation in particular, has become a major industry and is one of the fastest growing sectors of the world economy (Janic,

1999). Following an increased concern for the environment in general (Lutz, E. & Munasinghe, M. 1994; Willems, 1994; Jackson, 2010; EEA, 2009), pollution from airports and roads (EPA, 1995) has received more attention. It is also considered one of the contamination threats to soil according to the proposed EU Soil Framework Directive introduced by the European Commission in 2006 (COM (2006) 232; Tóth et al., 2008). In this chapter we will discuss various processes that need to be considered from the source to the recipient, which could be the groundwater or surface waters, and how these are affected by cold climate (winter frost). The source will be related to road network or runways, but can potentially be both a line source if no collection or surface drainage is in place, or can constitute a point source. In cases where road or runway runoff is collected in retention systems with subsequent infiltration into local soils or release to surface waters such as creeks or rivers it could potentially be a significant point source. Often there are no legal limitations on total amounts of salt that can be applied on roads and highways, hence there is no control of their release to the environment. Climate change may cause increased fluctuations about the freezing point which is a condition which increases the use of salts (French et al., 2010). Hence these chemicals may constitute a long term threat to soils in areas with frozen conditions in winter. The second process is infiltration into frozen, partially frozen or unfrozen soils, depending on state of soil and snow fall in late autumn/early winter. Frozen soils may create impermeable surfaces and highly affect hydrological conditions and in particular boundary conditions for unsaturated flow. Further, the flow and transport in the unsaturated zone is affected by soil physical and bio-geo-chemical heterogeneities and in cold climate, high temporal variability in degradation potential due to low temperatures during winter and snowmelt. If or when chemicals arrive at the groundwater level heterogeneous conditions continue to influence the fate of de-icing chemicals but the general mechanisms that apply have been widely documented elsewhere. In the next sections these processes and how they can be measured and modeled will be presented and we suggest how this knowledge can be used in planning of management strategies. The focus of this chapter is on processes near the surface and in the unsaturated zone. In the conclusions we discuss challenges still unresolved.

DE-ICING CHEMICALS

There are two types of de-icing chemicals; non-degradable inorganic salts and degradable organic compounds. Table 1 shows de-icing chemicals and their associated chemical oxygen demand in the case of complete degradation. For the inorganic salts, the only means of reduction is dilution.

Table 1: Most commonly used de-icing chemicals. [1)] Chemical oxygen demand (COD) mg/l based on standard stoichiometrical calculations assuming complete degradation [2)] assuming reduction to NH_4.

Group	**Name**	**Chemical formula**	**Chemical oxygen demand (COD[)1])**
Inorganic salts	Sodium Chloride	NaCl	0
	Magnesium Chloride	MgC_l2	0
	Potassium Chloride	KCl	0
Organic salts	Potassium Formate	KCOOH	0.35
	Sodium Formate	NaCOOH	0.35
	Calcium Magnesium Acetate	$CaMg(CH_3COO_)4$	4.28
	Sodium Magnesium Acetate	$NaMg(CH_3COO_)3$	3.21
	Potassium Acetate	KCH_3COO	1.07
	Sodium Acetate	$NaCH_3COO$	1.07
Others	Propylene Glycol	CH_3CHOCH_2OH	1.69
	Ethylene Glycol	$HOCH_2CH_2OH$	1.29
	Diethylene Glycol	$(HOCH_2CH_2)_2O$	
	Fructose/glucose/ Sodium Chloride	$C_6H_{12}O_6$/ $C_6H_{12}O_6$/NaCl	
	Urea	$(H_2N)_2CO$	2.1[32)]

INORGANIC SALTS

Some examples of total use of salts on Norwegian national roads per winter season are given in Table 2. The average salt consumption in 2009/2010 was 14 tons per km high priority roads.

Negative consequences of increased salt concentration along roads have been documented (e.g. Nystén and Suokko, 1998; Oberts et al., 2000). Howard & Maier (2007) simulated increased NaCl concentrations in groundwater due to urbanisation near Lake Ontario. According to their Visual Modflow simulations, maximum concentrations of 5000 mg NaCl/l were predicted near roads with an annual use of 20-250 tons NaCl per km road (dual to multiple lane motorways), and that stabilisation occurred after 700 years.The consequences for the environment can be groundwater of insufficient quality for drinking water. Direct or indirect release to lakes can give hypolimnetic conditions and prevention of biannual full circulation with reduced oxygen levels (Bækken et al., 2006). Release into creeks can give negative effects to biologic systems (Meland, 2010).

Table 2: Total consumption of road salts and sand on Norwegian national roads (the Norwegian public roads administration, www.vegvesenet.no)

Winter season	**NaCl (tons)**	**Sand(tons)**
2005/2006	166,000	435,000
2006/2007	137,000	386,000
2007/2008	160,000	484,000
2008/2009	200,000	443,000
2009/2010	201,000	417,000

ORGANIC DE-ICING CHEMICALS

Organic salts such as Potassium Formate, Potassium Acetate and Calcium Magnesium Acetate are used on runways and some roads (US EPA, Amundsen et al., 2008). These chemicals are more costly than the inorganic salts, but are used because they are degradable and less corrosive. Normally these chemicals are also released directly into the ground next to the road or runways.

Propylene glycol, Ethylene glycol and Diethylene glycol are the main constituents of de-icing chemicals used to keep planes and other vehicle surfaces free of ice. The de-icing at airport takes place on special de-icing platforms which normally collect the surplus chemicals with subsequent re-cycling or treatment in treatment plants (Øvstedal & Wejden, 2007). Estimations from Oslo airport, Gardermoen, suggest that 80% is collected at the de-icing platforms, about 10% is released to the local environment at take-off, and 10% leaves with the planes (Øvstedal & Wejden, 2007). Still many airports in the world do not collect the de-icing chemicals and they are released into the local environment (US EPA).

The treatment of water contaminated with organic de-icing chemicals is usually based on in-situ or off-situ aerobic and/or anaerobic degradation. The tested techniques vary from constructed wetlands, reed beds, constructed soil filters (Roseth and Bjørnstad, 1998; Roseth et al., 1998) natural soil profiles (French et al., 2001, Jaesche et al., 2006) to more conventional wastewater treatment plants (Rusten et al., 1999). The chemicals are supplied during the frozen season which may vary geographically but falls roughly between October to April. The objective of using de-icing chemicals is to reduce the freezing point of water, hence water will remain in a fluid state below 0°C and can infiltrate into the ground as long as there are open pores. Water containing de-icing chemicals may therefore infiltrate the ground prior to the main snow melting period. The preferential melt-out of de-icing chemicals was shown by French & van der Zee (1999) as also shown for other chemicals stored in snow (Johannesen and Henriksen, 1978). Although the organic de-icing chemicals are degradable and the top soil hosts large amounts of bacteria and fungi, the temperatures are negative or close to zero, and there is little degradation during this period. Half-lives of Propylene Glycol and Acetate under field conditions throughout melting period and into the summer ranged between 15-46 days (French et al., 2001). Degradation rates of other airport chemicals, as well as the increase of rates caused by optimizing the C:N:P ratio are reported in French et al. (2002). Although degradability of de-icing chemicals in general is positive for the environment, negative consequences can be anaerobic conditions. Field experiments (French et al., 2001) showed an increase in manganese concentration when propylene glycol and Potassium acetate was supplied to the ground surface even in the unsaturated zone, indicating oxygen limitation. On-going research (French et al., 2009) suggests that nitrate could increase remediation and improve redox conditions in local soils.

NON-INVASIVE VERSUS DESTRUCTIVE METHODS FOR SOIL CHARACTERISATION

The theoretical considerations concerning spatial variability, described in the next section requires that we have some knowledge about 1) the geostatistics of the hydrogeological properties, and 2) the scale of the contaminant source or plume relative to this variability.

Non-Uniform Infiltration

The boundary conditions influencing the flow and transport during snowmelt are characterized by ground frost and the formation of impermeable ice on the ground surface which redistributes melt water during the snowmelt period (Fig. 1). The ice-cover often inhibits infiltration in sediments with otherwise high infiltration capacity (sandy aquifers). As a result, a large amount of melt water collects in depressions or becomes surface runoff. According to Baker & Spaans (1997); Derby & Knighton (1997); Johnsson & Lundin (1991), infiltration during snowmelt often occurs as focused recharge in local depressions on the surface.

Monitoring water accumulation in snow and also the infiltration below the snow cover is a challenge. The traditional but laboursome way of collecting information about snow cover is to measure snow depths along a transect and estimate the snow water equivalent (SWE) by weighing the snow column. The advantage is that one gets a spatial coverage, but only for single times, also the method is destructive. Snow pillow measurements involves placing a logged scale beneath the snow cover, hence direct measurement of SWE above the snow pillow, this method prevents infiltration and is only representative of one location, also it may give wrong values when ice crusts are formed in the snow which reduces the weight load on the scale. Remote sensing using natural emissions of gamma rays and micro waves has become a common method for mapping snow storage over larger areas (Glynn et al., 1988; Durand et al., 2008; Schaffhauser et al., 2008), but also local measurements can be conducted using this principle (Campbell scientific; Bland et al., 1997).

A snow lysimeter is a method where meltwater is collected via drain pipes from trays below the snow cover, and volumes and quality measured. Hence destructive in the way that water is prevented from infiltrating the ground. Variations in release of water from a melting snowpack was documented byFrench & van der Zee (1999) by collecting meltwater from snow lysimeters placed beneath a melting snow cover. Despite a fairly uniform snow depth over the monitored area (a few square meters), the total melt volumes varied from nearly 0 to 200% of expected values based on the snow water equivalents measured prior to snowmelt. To further characterise the infiltration pattern in the soil,French & Binley (2004) installed electrodes near the surface and monitored changes in electrical resistivity of the soil volume, which indirectly reflected changes in water contents during snowmelt. Figure 2, shows the spatio-temporal variability of electrical resistivity observed in a horizontal plane during this experiment. This may cause higher velocities through the unsaturated zone than during evenly distributed infiltration on the surface, hence causing less than optimal conditions for degradation of pollutants. Preferential meltout of chemicals and subsequent redistribution of meltwater may also cause concentrations in soils next to the impermeable covers which are higher than expected from a pure mass-balance point of view. In addition to the temporary and variable surface conditions created by frost and snow, impermeable surfaces, membranes and other installations near the soil surface may highly affect the infiltration and flow pattern in the unsaturated zone next to roads and runways (e.g.Apul et al., 2007).

FIGURE 1: Surface conditions at Oslo airport, Gardermoen, near the end of snowmelt, showing patchiness of snow and ponding of melt water due to soil frost.

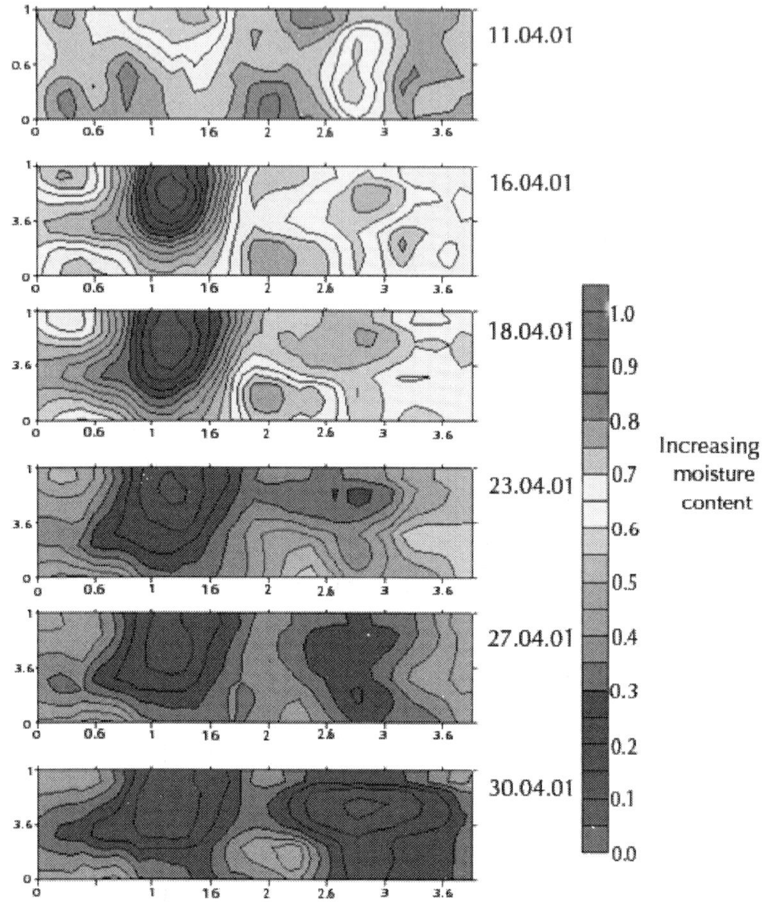

FIGURE 2: Variable infiltration, increased moisture content as blue areas, observed as reduced electrical resistivities near the surface of a soil below a melting snow cover (modified from French & Binley, 2004).

CHARACTERISATION OF SOIL HETEROGENEITY

Conventional techniques, such as soil sampling and core drillings provide more or less disturbed samples on which one can measure porosity, unsaturated hydraulic conductivity and Pf curve (if undisturbed) or grain size distribution and water content on disturbed samples. The Pf curve describes the relationship between soil suction and soil water content, this curve determines the unsaturated hydraulic conductivity (ref). Well established pedo-transfer functions such as Hazen's equation (ref.) can be used to calculate hydrogeological parameters such as the hydraulic conductivity. The samples can also be used for bio-geo-chemical characterisation interactions between contaminants and soil. A number of spatially collected samples can then be used to establish geostatistical properties of the different hydrogeological parameters (discussed in the next section). More recently advanced direct push technology, which provides opportunities for in-situ measurements e.g. of hydraulic conductivity by the use of specialised probes at the end of the direct push probe. The disadvantage of these methods alone is that they are destructive, time consuming, expensive; and do not give a continuous image of the subsurface. Deep geophysical exploration has been around since the beginning of the last century, and is a common method for geological characterization in oil exploration, mapping of lithostratigraphy, fracture patterns in bedrock and is described in several text books (e.g. Kearey & Brooks, 2002). In the last couple of decades geophysical techniques such as those described in Table 3have become more common for hydrogeological applications, for further reading see e.g. Regli, et al, 2002; Hubbard & Rubin, 2000, Kowalsky et al., 2001; Rea and Knight, 1998; Rubin & Hubbard, 2005;Veerecken et al., 2006.

As an example we discuss briefly the principle of the electrical resistivity method. The electrical properties of soils are a function of the soil type, water content, soil temperature and ion content of the soil water. Measurements of soil bulk electrical resistivity are most commonly conducted by placing a set of electrodes in the ground along a line on the surface or in vertical boreholes. By inserting a known current and measuring the resultant voltage consecutively over the set of electrodes, one can after an inversion of the collected data obtain an image of the distribution of electrical resistivities in the soil volume next to the electrodes (see e.g. Reynolds, 1997). A single measurement may reveal geological features of the subsurface, while the comparison of images taken at different times (time-lapse measurements) can help quantify spatial and temporal variability caused by changes in water (Daily et al., 1992) and ionic contents. The advantage of geophysical techniques over the more conventional and invasive techniques is that they are non-destructive and provide continuous images of the subsurface. The challenge of geophysical methods however is the ambiguity of their interpretation. The non-invasive geophysical methods map zones or layers of different physical characteristics (Table 3). The interpretation of such data requires that the data is run through an inversion code which basically "suggests" a likely distribution of the specific geophysical responses in a 1, 2 or 3D space. The results are optimised with respect to measurements conducted on the surface or in boreholes. Forward and inverse modelling of the system that is being studied is required for optimising the configuration of measurements, this technique can also be used in a stochastic framework in order to include uncertainty and coupling to soft and hard data for hydrogeological characterisation (Rubin and Hubbard, 2005). Another recent development to reduce the non-uniqueness of the interpretation is to combine different geophysical data sets collected at the same location and time through joint inversion (e.g. Gallado & Meju, 2004; Linde et al., 2006). However combining measurements from geophysical techniques and ground truthing (bio-geo-chemical data) is still required for more complete and accurate characterisation of spatial variability of hydrogeological parameters such as porosity, hydraulic conductivity and contaminant concentration.

TABLE 3: Geophysical methods for soil physical and hydrogeological characterization

Method	**Sensitive to**	**Procedure**	**Can represent**	**Example references**
Seismic refraction/ reflection	P-wave velocity and reflectivity	Sound wave, Sound source and set of geophones	Top of bedrock, lithology, water table, faults	Steeples, 2005;
Ground penetrating radar	Dielectric constants	Electromagnetic waves, Transmitter and receiver antennas,	Lithology, water table, water contents	Annan, 2005; Farmani et al., 2007

Electrical resistvity	Electrical resistivity	Electrical current injected and voltage measurements, usually electrodes in the ground	Lithology and zonation, water table, contaminant plumes, water contents	Binley & Kemna, 2005; Daniels et al., 2005;Kemna et al., 2006; Godio & Naldi, 2003
Electro-magnetic	Electrical resistivity	alternating or pulsed current through long wires or coils	Zonation, water table, contaminant plumes, water contents	Everett & Meju, 2005;
Induced polarisation	Chargeability	Same as for electrical resistivity, but non-polarisable electrodes	pore-fluid conductivity and grain-surface polarisation – can be used with ER data to separate lithological units	Binley and Kemna, 2005; Kemna et al., 2006; Vanhala et al., 1992
Self-potential	Streaming or Electrochemical potential	Passive voltage measurements, non-polarisable electrodes	Zones with concentrated flow, zones with high degradation	Naudet et al, 2003; Revil et al., 2006; Arora et al., 2007

CHARACTERISATION OF CONTAMINANT PLUMES

Solute transport, which is basic to natural attenuation processes, requires new measurement techniques to provide spatial distributions and internal spatial characteristics as current modelling of these phenomena is conditioned by availability of experimental data. Traditional monitoring techniques may neither capture the contaminant distribution nor their removal rates sufficiently (Aagaard et al., 2004). Subsurface characterization of contaminant distribution over large scales is challenging, since the contaminants may have moved erratically through the profile as illustrated in Figure 3, and point samples as provided with conventional sampling techniques (as discussed above) may not provide a representative measurement of the situation. The most common measurement technique for monitoring contaminant transport in the unsaturated zone is sampling of soil water through suction cups. These consist of a porous material such as ceramics or an inert material e.g. Teflon and quartz with a pore size small enough to ensure contact between the filter and the soil. An example of such a system is the experimental lysimeter trench at Moreppen near Oslo airport, Gardermoen, Norway (French et al., 1994). This experimental site has more than 100 Prenart suctions cups and various other soil physical measurements. It was constructed to conduct controlled experiments of transport of Propylene glycol and Potassium Acetate during snowmelt. Several studies were carried out at the same site in order to examine the hydrogeological properties in the unsaturated and saturated zones and the transport processes. Contaminants may, depending on their chemical properties, affect the geophysical signature of the soil. Salts will increase the electrical conductivity (EC) of the pore fluid, while hydrocarbons will have the opposite effect. The organic and inorganic de-icing salts will reduce the electrical resistivity of the soils, while Propylene glycol will not affect the electrical conductivity of the pore fluid. Electrical and electromagnetic methods are widely applied for soil mapping and detecting of contaminated plume. Over the last decade new geophysical methods such as induced polarisation (e.g. Godio and Naldi, 2003; Slater & Mansoor, 2006), electromagnetics, GPR, micro-sesimics and self-potential (Naudet et al., 2003; Arora et al, 2007) have been explored as methods for exploring contaminated sites. Low frequency electromagnetic (EM) methods are usually adopted for fast mapping and preliminary assessment of the aerial extent of the potentially contaminated land. A qualitative image of the soil mineralization, due to degradation of hydrocarbons, could be inferred by integration of resistivity and induced polarisation data (e.g. Godio and Naldi, 2003, Slater et al. 2006). Electrical Resistivity Tomography (ERT) is a powerful tool for investigating pore fluid properties (Olsen at al., 1999; Kemna et al., 2000; Depountis et al., 2001; Damanesco and Fratta, 2006;) as shown in laboratory experiments (Comina et al., 2005) and for solute transport in undisturbed soil columns (Binley et al., 1996) and field sites (Slater et al., 2000; French et al., 2002, Binley et al. 2005). How to estimate hydrogeophysical parameter is still one of the major challenges, state-of the art knowledge is described by Linde et al., (2006). Another challenge for combined interpretation of geophysical and point measurements is that the support scale of different methods varies; hence a statistical framework is required for joint interpretation.

MODELLING IMPLICATIONS

As evidence shows, the subsoil is in general heterogeneous (or spatially variable), and often this heterogeneity is partly irregular. This irregular variation has been the motivation to consider soil as an intrinsically random material, i.e., as a material that can only be described statistically. This assumption has resulted in a large body of literature (Bellin et al., 1993; Dagan, 1997; Keijzer et al., 1999; Janssen et al., 2006; Cirpka, O.A., P.K. Kitanidis, 2000; Fiori et al., 2002), that is still actively being developed and is quite mathematically inclined: stochastic groundwater hydrology and contaminant hydrology. For the present purpose, it is not useful to emphasize the mathematics, although some of it cannot be avoided. Rather, emphasis is given to illustrate the effects of heterogeneity and to give an impression of what that means for decision making.

Stochastic Hydrology

The mentioned stochastic approach to contaminant hydrology (that focuses on the transport behaviour and fate of soil and groundwater contaminants) implies that particular properties are considered to be random space functions (RSF): they vary irregularly in the horizontal plane and with depth. However, they do not do so completely randomly: due to the genesis of layering, and soil horizons, patterns of large and small values can be observed. These patterns are described by autocovariance functions. Examples of strongly variable properties are the hydraulic conductivity, and for the water unsaturated zone also the soil water retention function (Van der Zee & Boesten, 1991). Much research has established that the hydraulic conductivity, and the scaling parameter in this function that also is used to describe water retention, are well described with a lognormal distribution. Also for several important properties that control the adsorption behaviour of contaminants, the lognormal distribution appeared to be reasonable (Van der Zee & Van Riemsdijk, 1987, 1988; Boekhold & Van der Zee, 1991).

Randomness can have a large effect on flow and transport (Rode et al., 2010). Basically, this effect is (i) a more irregular transport behaviour of the contaminants, and (ii) a larger uncertainty about this behaviour.

To give an impression of how heterogeneity affects the transport pattern, Figure 3 shows the leaching behaviour of both inert chloride and degrading de-icing chemical at Oslo Airport, Gardermoen (French et al., 2001). We observe a very irregular 'plume' of contaminant. A problem with such complex patterns is that they are difficult to communicate, other than by sending a picture.

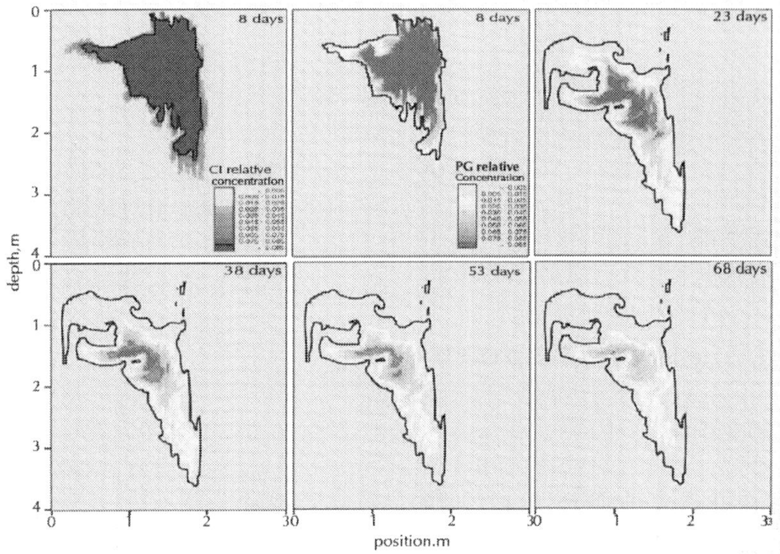

FIGURE 3: The erratic pattern of chloride and propylene glycol transport in a heterogeneous soil representative of the Moreppen site at Gardermoen (French et al., 2001).

In addition, we have to deal with uncertainty, because the pattern if the contaminant were released on a spot a few meters away might look similarly irregular but not exactly the same! It might be slightly deeper or shallower, moved to the left or right, and have another shape. Whereas for contaminated sites, it may be meaningful to know the exact pattern, for making predictions, the precise shape of the plume becomes less important: a calculation for another spot would look always more or less different. For this reason, we need more robust measures to quantify the transport behaviour, than the pictures of Figure 3.

Spatial Moments

To this aim, we explain the moment theory, with which such quantification is feasible at different levels of detail. Using this moment theory, it is feasible to illustrate the profound impact of spatial variability on contaminant transport in terms of pollution front, breakthrough time, and of course remediation efficiency.

If soil and aquifer properties are spatially variable, two limiting situations may arise. These two limiting situations are related with those of point source pollution and of diffuse source pollution, respectively. The point source pollution problem can be defined as the situation where the source is relatively small (or of similar order of magnitude) compared with the scale of (random) heterogeneity. In that case, the source of contaminants might be visualized as being in a spot of large or of small hydraulic conductivity. In the first case, contaminants may readily spread in the environment, whereas in the latter case, this occurs slowly. For the point source pollution event, it is difficult to predict whether the source will be at in contact with the fast or the slow route of transport, which leads to high uncertainty. In the diffuse source pollution case, contaminants enter the soil or aquifer over a large surface, and both fast and slow transport routes are 'sampled'. In that case, uncertainty is much smaller, and heterogeneity leads to large spatial variability of fluxes. This distinction of two limiting situations is illustrative, but in view of the nested scales of heterogeneity of soil and aquifers, many real situations will be somewhere in between.

Figure 4: Photo of a heterogeneous soil profile.

It is clear from Fig. 4 that the thickness and intensity of soil horizons varies in the horizontal plane. As soil genesis processes involve longer time periods, spatial variability must be due to persistent differences in soil composition and do not necessarily result from differences in 'input' at the soil surface.

For the two limiting cases mentioned above, the 'diffuse' pollution case is called an ergodic situation, because 'all' heterogeneity is sampled simultaneously. This leads to a distributed answer, such as a spatially variable contaminant plume or breakthrough curve. If these answers are measured or calculated at another place, they do not significantly change. Hence uncertainty is limited.

However, the distributed answer is complicated and difficult to communicate to others who cannot see the picture. Moreover, it is commonly too detailed for e.g. management decisions. Therefore, it is appropriate to consider robust measures of the results, which focus on the major issues. Useful examples of such robust measures are found in the theory of moments. We will give the definition of these moments in terms of spatial moments (representing the situation at one instant in time), but temporal moments (at one particular location, plane or volume) are equally feasible. The spatial moments of a property P, where the space coordinate is x, are given by

The zeroth' moment, or the mass of the distribution:

$$M_0 = \int_{-\infty}^{\infty} f_p dP \tag{1}$$

The first moment, or the mean of the distribution:

$$M_1 = \frac{1}{M_0} \int_{-\infty}^{\infty} P f_p dP \tag{2}$$

The second central moment, or the variance of the distribution:

$$M_2^c = \frac{1}{M_0} \int_{-\infty}^{\infty} P(P - M_1)^2 f_p dP \qquad (3)$$

For a contaminant plume as shown in Figure 3, the zeroth' moment is equal to the amount of dissolved chemical in the contaminant plume, if P is equal to the position x, and the function fx represents the spatial distribution of the contaminant concentration multiplied with the water-filled porosity. The first moment represents the mean position (in direction x) of the contaminant distribution and the second central moment represents the width of the zone over which the concentration distribution occurs. Whereas we illustrated the moments where P is equal to position, many other properties can be chosen. Which to choose depends on the primary interest, but examples are the concentration or mass of contaminant, the quantity or fraction of contaminant leached beyond a reference plane, such as groundwater level and so on. In principle, the transport problem for de-icing chemicals is a very complicated one, even in one direction such as depth. The reason for that is that a complex of interactions is affecting this transport. Examples are infiltration conditions, ad/desorption, microbial degradation according to different kinetics (1st, Monod, 0th order), and a host of different redox-sensitive components that may affect the degradation rate (Beltman et al., 1996; Keijzer et al., 1998) may occur, depending on ambient conditions as well as the influx of degrading substrate (continuous, instantaneous). To give an impression of the complexity, we provide graphs in Figure 5, that reveal that the concentration pattern may be completely different from what we are used to, and that the transport process is quite irregular in space.

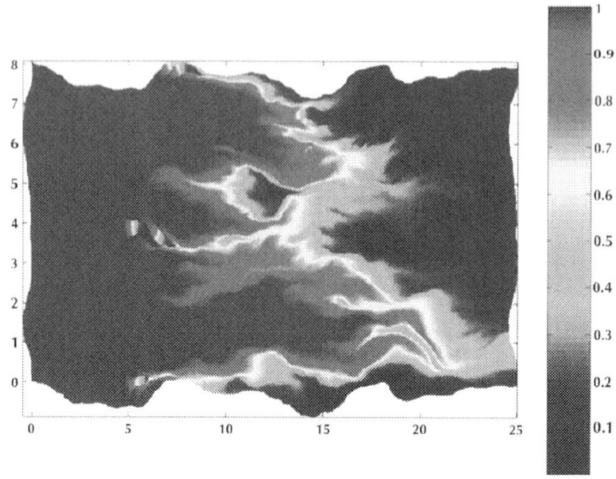

Figure 5: Computationally determined map of concentrations of an electronacceptor (EA) such as oxygen or nitrate with scaled concentrations ranging from 0 to 1. The EA enters the aquifer from the left and moves towards the right, and is used by micro-organisms to degrade initially resident gasoline that contaminates the aquifer. Computationally, this situation is quite demanding to solve, and a streamline dependent spatial discretization was needed to suppress numerical dispersion (Janssen et al., 2006).

Figure 6 gives a representation of the behaviour of Figure 5, in terms of concentration distributions in the longitudinal direction. This result implies that the three linked concentration curves for initially resident degradable contaminant G (gasoline, or kerosene, for instance), an incoming (from the left) electron acceptor (EA such as oxygen) C, and resident, growing and decaying bacterial mass M, show patterns that for each time are directly related with each other. What is complicated is that the electron acceptor concentration distribution for all three times is different, and that between time 1 and 2 the injected electron acceptor seems to withdraw to the left again. This seemingly illogical result is purely the consequence of the increasing microbial mass when both substrate (G) and e.g. oxygen are present, as then, the consumption of G suddenly increases and its concentration decreases. As Figure 5 reveals, spatial variability of the hydraulic conductivity may strongly affect the transport of electron acceptor and the related degradation of resident contaminant. Erratic patterns may develop, that are difficult to communicate.

Whereas the complexity of the transport problem is profound, the strength of identifying the right robust features may inspire to equally profound simplifications. Considering the downward transport of a degrading contaminant towards the groundwater level, the mass balance equation for linearly adsorption and 1st order degradation reads:

$$\rho\frac{\partial q}{\partial t}+\theta\frac{\partial c}{\partial t}=\theta D\frac{\partial^2 c}{\partial x^2}-\theta v\frac{\partial c}{\partial x}-k(\theta c+\rho q) \tag{4}$$

Where q is adsorbed amount, c is concentration, k is degradation rate, v is pore velocity, D is dispersion coefficient, θ is volumetric water fraction, ρ is density.

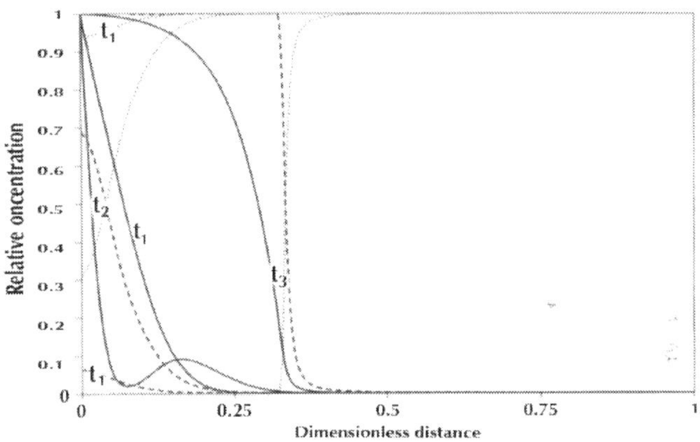

FIGURE 6: Concentration distributions for electron acceptor (solid), contaminant (dotted), and microbial biomass, for times increasing in the direction of t_1 to t_3. Concentrations scaled with regard to their maximum concentration.

In essence, according to this transport equation, adsorption/desorption is instantaneous, and the degradation occurs (according to the last term) both in the liquid and in the solid phases. In terms of system properties, physical properties such as the hydraulic conductivity and water retention, control the development in time of the volumetric water fraction (θ) and flow velocity (v). Biogeochemical properties of importance are those related with the adsorption equation, s(c), e.g. affinity coefficient and a sorption maximum, and those related with the degradation rate, e.g. Monod kinetics parameters that describe the degradation rate as a function of the concentrations of contaminant, of electron acceptors or reductors, and microbial activity. Commonly, degradable contaminants are assumed to degrade according to first order kinetics. For the given transport equation, the transport and leaching process can be simulated with a computer model such as PEARL (Leistra et al., 2000). The PEARL model considers one-dimensional transient flow of water in the soil (vadose zone) profile, which is linked to the convection dispersion equation (CDE) for transport of chemicals. Besides that the model has modules for crop growth and energy and water interactions with the atmosphere, it is also possible to account for (monocomponent) adsorption/desorption and first order degradation. The PEARL model(see Beltman et al., 2008) is used in pesticide screening in the EU, for which the FOCUS scenarios have been developed ((FOCUS, 2000).). These scenarios harmonize the soil type, climate and weather variations within the EU for which the assessment of pesticide leaching to groundwater is evaluated, with several aims, among which the need to consider a limited number of modeling permutations, while still enabling comparison of results EU-wide.

IMPLICATIONS FOR MANAGEMENT STRATEGIES

Simplified Modeling Approach

For decision making, we propose to first make an approximate assessment based on several simplifications. Thus, diffusion-dispersion is considered to be of secondary importance and flow is assumed to be steady state (as a first step). For groundwater quality protection, particularly the leached fraction of applied or spilled contaminant is of interest, because that quantity will control the concentration in the annual recharge of the topmost aquifer. For a particular year, both the leached fraction and the recharge (precipitation minus evapotranspiration) may be difficult to predict, which an example of lack of ergodicity is in time. However, if the interest is primarily for long term predictions as is the case with de-icing chemicals applied every winter, rather than for incidental spill events, then the uncertainty becomes considerably smaller. Regular leaching leads in that case to repetitive concentration jumps in the upper groundwater aquifer, and whether these jumps lead to a gradual building up of concentrations in groundwater, can be analysed according to the method of Beltman et al. (1996) and Van der Zee et al. (2010).

In those papers, we showed how the regular leaching of contaminant can result in a building up of concentrations in groundwater, depending on the capacity of the soil to degrade the contaminant: Figure 7. Whereas Beltman et al. (1996) focused on degradation and transport in an aquifer, Van der Zee et al. (2010) considered leaching from a mixed reservoir analogous of the unsaturated soil. In essence, the mathematical formulation does not depend on whether the first order kinetics of the contaminant transport and removal are due to transport processes or to degradation. Hence, in analogy to the approach of Van der Zee et al. (2010), we can consider an unsaturated soil volume, that loses contaminant to the groundwater. The quantity that is lost, we consider below, but if the lost quantity is equal to M and the volume of water in the unsaturated soil zone equals V, then

$$C = \frac{M}{V}\frac{\left[1-\left(\exp\left(-\frac{j\Delta t}{V}\right)\right)^{n}\right]}{1-(\exp(-\frac{j\Delta t}{V}} \tag{5}$$

For the maximum concentrations of the resulting saw tooth pattern of concentrations, where j is the water flux leaching from the unsaturated soil, n is the number of applications of the contaminant, and the time step is for one year. Observe that the minimum concentrations of this pattern are given by C-M/V.

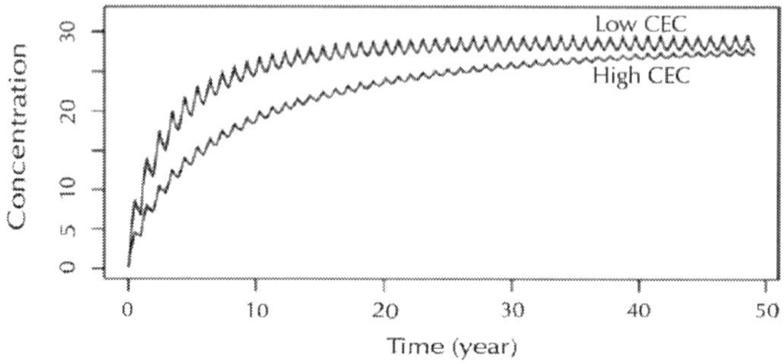

FIGURE 7: Saw tooth patterns of concentration for an application of contaminant every year for two different values of CEC (cation exchange capacity), which controls concentration buffering. From: Van der Zee et al. (2010).

For a homogeneous soil, the leached fraction, *F*, according to the Convection Dispersion Equation is given by

$$F = \exp\left[\frac{-\mu RL}{v}\right]; R = 1 + \frac{\rho}{\theta} Ks \tag{6}$$

If dispersion is ignored, where R is the retardation factor and Ks is the sorption coefficient for linear sorption. In this solution, we recognize the soil physical and atmospheric forcing controlled properties (θ, v), and the contaminant specific properties (Ks, and degradation rate parameter μ). Hence, a leaching vulnerability assessment system for the applicability ofde-icing chemicals could involve a water flow and tracer leaching investigation, followed by a contaminant specific leaching investigation. In real soil systems, the various parameters of equation (6) vary in space and time, and this may have a large effect on the leached fraction. Therefore, Van der Zee & Boesten (1991) simulated leaching for the heterogeneous case. It appears that if there is a weak spot in the soil where leaching is relatively large and fast, then this one spot can contaminate a large volume of water up to the level of the water quality standard even if elsewhere no leaching of contaminant occurs. For moderate variability of soil properties, the leaching of contaminant in such a heterogeneous environment still behaves as 'first order degradation and transport', as described by equation (5), but the constants in this equation should be appropriately averaged.

How this averaging should be done, focused on the leached fraction, is explained by Van der Zee & Boesten (1991).

Aimed at the problem of contamination with de-icing chemicals, it appears feasible to approach surface runoff also from a partly stochastic approach. Therefore, some first indication of ways to deal with irregular soil surfaces in the case of overland flow are presented using the framework of Appels et al. (2011). Using these approaches, it appears to be possible to give a stochastic analysis of the in situ aquifer bioremediation of degradable contaminants such as Propylene Glycol. Besides illustrating this aspect, many problems and unresolved questions are outstanding. For instance, mostly soil and groundwater formations are assumed to conform to Gaussian spatial structures, but in reality, more complicated structures that are distinctly non-Gaussian, may be more realistic.

Remediation in Heterogeneous Soils

Because transport of nonlinear reacting contaminants in spatially variable soils is a very active field of research, many problems are unresolved, yet managers need to make decisions, for instance concerning choice of remediation technique. For this reason, it is appropriate to indicate in what respect management decisions need to take into account this complexity. For instance, the applicability of the still recommended (e.g. by USA, EPA) air sparging as a methodology to purify groundwater that is contaminated with organic biodegradable or volatile contaminants can be judged well on the basis of the physics of this technique.

In air sparging, air is injected below the phreatic groundwater level, with the purpose to strip the aquifer from volatile contaminants, but also to increase the oxygen content in those strata as certain organic contaminants require it to become degraded. Much effort has been devoted to experimentally investigate how air will flow through the aquifer (upwards, as air is much less dense than water): in a narrow, vertical funnel or in a gradually upward broadening cone. The latter occurs if the aquifer material is sufficiently fine textured sand, but finer, loamy or clayey textures cannot well be sparged in view of their air entry values. Engineers were particularly interested in the so called radius of influence (ROI), which is the radius of the supposedly circular cross section of the cone through which air flows at the phreatic groundwater level. In addition, air saturations in this cone are of interest. A mathematical analysis has been worked out by Van Dijke et al. (1995) and was later refined by Philip (1998). In both cases, relatively simple solutions were obtained, but in both cases for a single cone and for a homogeneous aquifer. In a refinement, Van Dijke & Van der Zee (1998) considered a two layered medium, i.e., air was injected below the groundwater level, in a coarse layer that is overlain by a slightly finer textured layer. The contrast in physical properties of the two layers was very small (hydraulic conductivity differed only by a factor √2, i.e., hardly measurable). Even with such a small contrast in properties, the effect on the sparged area was major, as can be seen from Figure 8. It appears that at the interface between the two layers, the cone rapidly expands to larger radial distances away in a horizontal direction from the injection point, and air saturations rapidly decrease. Important is to note, that these calculations were done for an axially symmetric situation. The practical impact of these results is the following: if the air flow pattern is so sensitive to slight spatial (vertical) variability of the subsoil, for real subsoils with intricate spatial patterns in 3D, this must also be expected and (in view of variability) in an enhanced way. The air flow will be along a path of minimal resistance, and along this path, almost all flow will concentrate. Therefore, although injected air will oxygenate the subsoil, this will be a very inefficient way to do so, as flow is excessively preferential. Moreover, it cannot be well steered, to get the air to flow to a particular part of the subsoil, e.g. the contaminated part. In reality, the flow may follow many possible paths around a contaminated volume.

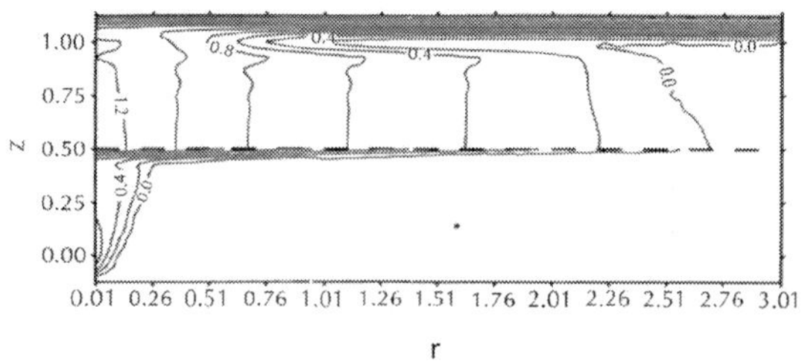

Figure 8: ISO-concentration lines of air saturation for a two-layered medium with layer interface at dimensionless depth Z. Dimensionless radial distance is r (Van Dijke and Van der Zee, 1998).

In more general terms, spatiotemporal variability leads to uncertainty and therefore some suggestions can be given regarding soil/water contamination decision making under uncertainty, focused at the de-icing chemical contamination context.

Guidelines for decision making

Many contaminated sites and polluting surface activities rely on natural or manipulated remediation to remove pollutants in the soil. In order for this approach to be acceptable it is important to have control of: (1) the pollution source, (2) the behaviour of pollutants in soils as influenced by spatial variability of the subsurface, (3) their removal rates (adsorption, degradation, precipitation) and how these can be optimised, and (4) long term effects on the soil system. In the above sections some of these aspects related to contaminants from winter maintenance in cold climate have been discussed, as well as different methods for subsurface characterisation and contaminant transport monitoring. The importance of dealing with heterogeneities in modelling strategy has also been dealt with.

Legal and Political Incentives

How contaminants are dealt with at contaminated sites in general, is not only determined by whether the scientific community can overcome present challenges as those discussed in the previous sections, but also within which political and legal framework contaminated site managers work under. Whether the optimal strategies for monitoring and cleanup are chosen depends on the following:

- A correct conceptual understanding of the processes which determine flow and transport of the contaminants, which is the prerequisite for correct choice of modelling tool to assist management at the site.
- The scientific knowledge, do we have the right methods and understanding of the processes to be able to deal with the contamination in a good way? which implies
- Good methods exists for mapping and monitoring surface and subsurface heterogeneities, at present a combination of

destructive and non-invasive techniques are recommended in combination with transport modelling.

- Robust relationships exist between geophysical properties and hydrogeological properties and sufficient resolution of survey at the site.
- Robust relationships exist between geophysical properties and contaminant situation in the soil, and sufficient resolution.
- A sound method for coupling of different methods destructive and non-invasive methods, for quantitative description of contamination levels and remediation efficiency.
- Appropriate remediation techniques for heterogeneous systems exist.
- Managers' knowledge about the system they are in charge of, the quality depends on:
- Size of company, in view of available resources made available for such purposes.
- Type of contaminant – available information about reactive properties
- Company regulations and policies
- The authorities knowledge about contaminants, survey possibilities, modelling and risk factors, as well as their capacity to follow up contaminated sites and validate that accepted procedures are being followed.
- Legal international and national framework regulating expectations to contaminated sites.

Cold Climate Implications

Scientifically many processes are still not well described for low and freezing temperatures, such as physical conditions in freezing soil and how to model these, especially the combination of de-icing chemicals which lower the freezing point in a soil system near 0°C is complicated. Degradation and biological dynamics as a function of low temperature is generally not well described. Instrumentation may work poorly under frozen conditions, good methods for measuring the total water content in a frozen soil are for instance still lacking. What scale of heterogeneity is relevant during snow melting conditions compared to the thawed situation? Cold climate conditions add extra complications to the contaminant site management, for one; the need for use of de-icing chemicals. Since the use is generally not regulated, there is a great potential for reduced environmental impacts by implementing such limitations. At Oslo airport, Gardermoen, only a certain amount of de-icing chemicals is allowed per winter season. De-icing chemicals that are highly concentrated such as at the de-icing platforms are collected, hence reducing the amount released to the green areas. Since soil temperature has a large effect on surface infiltration patterns, this should be monitored as part of the regular monitoring programme along roads and at airports. Strategies to deal with both frozen and non-frozen conditions during snowmelt must exist, e.g. pump and store contaminated melt water in local depressions for later infiltration when conditions are more optimal, e.g. in summer.

CONCLUSION

This chapter gives an overview of properties of de-icing chemicals and the processes that affect their fate in the unsaturated zone. Inorganic salt concentrations can only be reduced by dilution and is only discussed briefly here, while degrading organic de-icing chemicals are the main focus. Conventional methods of soil and soil water samples are still required for interpretation of emerging technologies such as different geophysical techniques. The importance of heterogeneity of the infiltration process and unsaturated flow is emphasised. The method of spatial moments is an efficient way of describing average movement of contaminants in the unsaturated zone. An ergodic situation, i.e. when the size of the contaminant source is much larger than the scale of soil heterogeneity, makes average assumptions more certain while small plumes in a system of large variability causes great uncertainty. Geophysical methods can be used in time lapse mode to obtain a more continuous impression of the flow and transport processes and is recommended in combination with point measurement for quantification of absolute levels of contaminants. Flow and transport modelling within a stochastic framework is another important complementary technique for optimised management strategies at contaminated sites.

Applied Hydrogeophysics Nato Science Series: IV: Earth and Environmental Sciences, 71Springer, 117-160

60. A. Kemna, J. Vanderborght, H. Hardelauf, H. Vereecken, 2004Quantitative imaging of 3D solute transport using 2D time-lapse ERT: a synthetic feasibility study. Proceedings of the Symposium on the Application of Geophysics to Engineering and Environmental Problems / Environmental and Engineering Geophysical Society, Colorado Springs,- S. 342353
61. M. B. Kowalsky, P. Dietrich, G. Teutsch, Y. Rubin, 2001Forward modeling of ground-penetrating radar data using digitized outcrop images and multiple scenarios of water saturation, WATER Water Res. Res. (6): 1615-1625,
62. M. Leistra, A. M. A. van der Linden, J. J. T. I. Boesten, A. Tiktak, Berg. F. van den, 2000PEARL model for pesticide behaviour and emissions in soil-plant systems. Description of processes. Alterra report 13, RIVM report 711401009, Alterra, Wageningen, 107 pp.
63. N. Linde, A. Binley, Pedersen. Tryggvason, A. Revil, 2006Improved hydrogeophysical characterisation using jont inversion of cross-hole electrical resistance and ground-penetrating radar traveltime data, Water Resour. Res.,42W12404, doi:10.1029/2006WR005131
64. E. Lutz, M. Munasinghe, 1994Integration of environmental concerns into economic analyses of projects and policies in an operational context, Ecological Economics1013746
65. Kay. A. A. Mac, P. M. Gschwend, 2001Enhanced concentrations of PAHs in groundwater at a coal tar site. Environmental Science and Technology, 35 (7), 1320-1328.
66. N. Mansoor, L. D. Slater, 2007On the relationship between iron concentration and induced polarization in marsh soils. Geophysics, 72, A1A5.
67. S. Meland, 2010Ecotoxicological effects of highway and tunnel wash water runoff, PhD thesis, Norwegian University of LifeSciences,
68. K. Metselaar, Lier. Q. De Jong van, 2007The Shape of the Transpiration Reduction Function under Plant Water Stress. Vadose Zone Journal, 6124139

69. V. Naudet, A. Revil, J. Bottero, J.-Y.,2003Relationship between self-potential (SP) signals and redox conditions in contaminated groundwater. Geophys. Res. Letts., 30 (21), 2091, doi:10.1029/2003GL018096.
70. T. Nystén, T. Suokko, 1998Transport processes of road salt in Quaternary formations. Deicing and dustbinding- risk to aquifers.Symposium, Helsinki, Finland, Oct 1416NHP Report 439-52110-348-50900-0267
71. G. L. Oberts, J. Marsalek, Viklander. M. og, 2000Review of water quality impacts of winter operation of urban drainage. Water Qual. Res. J. Canada 354781808
72. P.A.Olsen,A.Binley,S.Henry-Poulter,W.Tych,1999Characterizing solute transport in undisturbed soil cores using electrical and X-ray tomographic methods. Hydrol. Process., 13211221
73. J. Øvstedal, B. Wejden, 2007Dispersion of de-icing chemicals to the areas along the runways at Oslo Airport Gardermoen, SAE International, 2007-01-3351
74. J. R. Philip, 1998Full and boundary-layer solutions of the steady air sparging problem, J. Contaminant hydrology, 33, 3-4, 337-345
75. J. Rea, R. Knight, 1998Geostatistical analysis of ground-penetrating radar data: A means of describing spatial variation in the subsurface, Water Resources Research, 34329339
76. C. Regli, P. Huggenberger, M. Rauber, 2002Interpretation of drill core and georadar data of coarse gravel deposits. Journal of Hydrology, 255234252
77. A. Revil, K. Titov, C. Doussan, L. Vincenzo, 2006Applications of the Self-potential method to hydrological problems, in Veerecken H., Binley, A., Cassiani, G., Revil, A., Titov, K. 2006 Applied Hydrogeophysics Nato Science Series: IV: Earth and Environmental Sciences, 71Springer, 255-292
78. J. Reynolds, 1997An Introduction to Applied and Environmental Geophysics. Wiley, New York, 796 pp.
79. M. Rode, G. Arhonditsis, D. Balin, T. Kebede, V. Krysanova, A. Van Griensven, A. , S. E. A. T. M. , van der Zee (2010New challenges in integrated water quality modelling, Hydrological processes, DOI:10.1002/hyp.7766,

80. R. Roseth, H. Bjørnstad, 1998Airport stormwater treatment in constructed soil filters-degradation of a new runway deicier; AVIFORM L50TM, in T. Nysten and T. Suokko (eds.), Deicing and Dustbinding- Risk to Aquifers, Heslinki, Finland, 14-10-1998, Nordic Hydrological Programme Report 43145154

81. R. Roseth, H. Bjørnstad, P. Kraft, B. Warner, 1998Airport stormwater treatment in constructed soil filters- a comparative study of aircraft deiciers, in T. Nysten and T. Suokko (eds.), Deicing and Dustbinding- Risk to Aquifers, Heslinki, Finland, 14-10-1998, Nordic Hydrological Programme Report 43155164

82. Y. Rubin, S. Hubbard, 2005Hydrogeophysics, Water Science and technology Library, 50, Springer, The Netherlands, 523 pp.

83. B. Rusten, F. Wessman, I. Tranum, 1999Success with moving beds, Water Resources Bulletin April, 2325

84. A. Schaffhauser, M. Adams, R. Fromm, P. Jörg, G. Luzi, L. Noferini, R. Sailer, 2008Remote sensing based retrieval of snow cover properties, Cold Regions Science and Technology, 543164175

85. L. D. Slater, N. Mansoor, 2006The dependence of induced polarization on natural iron concentration in wetland soils. AGU conf.

86. L. Slater, A. Binley, W. Daily, R. Johnson, 2000Cross-hole electrical imaging of a controlled saline tracer injection, J.Applied Geophysics, 4485102

87. D. W. . Steeples, 2005Shallow seismic methods, in Rubin, Y. and Hubbard, S. (eds) Hydrogeophysics, Water Science and technology Library, 50, Springer, The Netherlands, 523 pp.

88. G. Tóth, L. Montanarella, E. Rusco, (eds, 2008Threats to Soil Quality in Europe, EUR 23438 EN- 2008;http://eusoils.jrc.ec.europa.eu/esdb_archive/eusoils_docs/other/EUR23438.pdf

89. S. E. A. T. M. van der Zee, J. J. T. I. Boesten, 1991Effects of soil heterogeneity on pesticide leaching to groundwater, Water Resour. Res., 27(12), 3051- 3063

90. Van der Zee, S.E.A.T.M., and W.H. van Riemsdijk1987Transport of reactive solute in spatially variable soil systems, Water Resour. Res., 2320592069

91. S. E. A. T. M. Van der Zee, M. Nederlof, W. H. van Riemsdijk, 1988Spatial variability of phosphate adsorption parameters, J. Environ. Qual., 17682688

92. S. E. A. T. M. Van der Zee, S. H. H. Shah, C. G. R. Van Uffelen, P. A. C. Raats, Ferro. N. Dal, 2010Soil sodicity as a result of periodical drought. Agricultural Water Management 974149doi:10.1016/j.agwat.2009.08.009

93. M. I. J. Van Dijke, S. E. A. T. M. , van der Zee (1998Modeling of air sparging in a layered soil: numerical and analytical approximations, Water Resour. Res., 34(4), 341-358

94. Van Dijke, M.I.J., S.E.A.T.M. van der Zee, and C.J. van Duijn1995Multiphase flow modeling of air sparging, Adv. Water Resour., 18319333

95. H. Vanhala, H. Soininen, I. Kukkonen, 1992Detecting organic chemical contaminants by spectral-induced polarization method in glacial till environment. Geophysics, 5710141017

96. H. Veerecken, A. Binley, G. Cassiani, A. Revil, K. Titov, 2006Applied Hydrogeophysics Nato Science Series: IV: Earth and Environmental Scien., 71Springer, 383 pp

97. H. G. M. Willems, 1994Turning environmental concerns into environmental care in the chemical industry in The Netherlands,

98. Yong R. N. and Hoppe E. J. (1989). Application of electric polarization to contaminant detection in soils. Can. Geotech. J./ Rev. can. geotech. 26, 536-550

Chapter 2

Effect of the Presence of Nonionic Surfactant Brij35 on the Mobility of Metribuzin in Soil

Eman M. ElSayed and Shiv O. Prasher

Department of Bioresource Engineering, McGill University, Ste. Anne de Bellevue, Quebec H9X 3V9, Canada

ABSTRACT

Given the water scarcity becoming endemic to a large portion of the globe, arid region irrigation has resorted to the use of treated, partially treated, or even untreated wastewaters. Such waters contain a number of pollutants, including surfactants. Applied to agricultural lands, these surfactants could affect the fate and transport of other chemicals in the soil, particularly pesticides. A field lysimeter study was undertaken to investigate the effect of nonionic surfactant, Brij35, on the in-soil fate and transport of a commonly used herbicide, metribuzin [4-amino-6-*tert*-butyl-3-(methylthio)-1, 2, 4-triazin-5(4*H*)-one]. Nine PVC lysimeters, 1.0 m long × 0.45 m diameter, were packed with a sandy soil to a bulk density of 1.35 mg m^{-3}. Antibiotic-free cattle manure was applied (10 mg ha^{-1}) at the surface of the lysimeters. Metribuzin was then applied to the soil surface of all lysimeters at a rate of 1.00 kg a.i. ha^{-1}. Each of three aqueous Brij35 solutions, 0, 0.5 and 5 mg L^{-1} (*i.e.*, "good", "poor" and "very poor" quality irrigation water) were each applied to the lysimeters in triplicate. Analysis for metribuzin residues in samples of both soil and leachate, collected over a 90-day period, showed the surfactant Brij35 to have increased the mobility of metribuzin in soil, indicating that continued use of poor quality water could influence pesticide transport in agricultural soils, and increase the risk of groundwater contamination.

INTRODUCTION

Water is a very complex and dynamic resource and although the estimated amount of the available water on the planet is 140 billion km^3, only 3% is fresh water [1]. Due to the distribution of fresh water resources around the globe, less than 1% of these resources are available for human use. This, in turn, leads to water shortage problems in many regions of the world, particularly developing countries. Water scarcity combined with poverty is forcing the reuse of untreated or partially-treated wastewater for irrigation in order to cope with the high food demands of growing populations [2]. Consequently, the risk of bacterial infections, roundworms and diarrheal diseases are much greater [3]. A recent survey revealed that the 46 countries, representing 75% of the world's irrigated land, use polluted water for irrigation [4]. Globally, an estimated 20 million ha of land are irrigated with wastewater, and this area is expected to markedly increase in the next few years [5].

Wastewater, originating from the discharge of municipal and industrial effluents, contains a number of inorganic and organic substances. Among organic contaminants of wastewater and sludge, surfactants are often detected at the highest concentrations [6,7,8]. Non-ionic surfactants concentrations ranging from 10^2 to 10^3 mg L^{-1} have been detected in wastewater [9]. Non-ionic surfactants such as alcohol ethoxylates (AEs), alkylphenol ethoxylates (APE) are used extensively as leather degreasing agents and in other textile industry applications. Consequently, non-ionic surfactants have been identified as major contaminants in untreated wastewater from such industries [10,11]. Owing a non-dissociable hydrophilic group (e.g., alcohol, phenol, ether, ester, or amide), non-ionic surfactantsdo not ionize in water. Although in the wastewater treatment plants, APEs degrade into metabolites such as nonylphenol ethoxylates (NPEs), treatment plants are not adequately efficient to remove the metabolites such as environmentally damaging nonylphenols (NPs) [12,13]. Consequently, nonylphenols (NPs) were detected at high concentrations ranging from 1.8 to 25 g L^{-1} in two rivers in Switzerland [14]. Further, studies showed that the continuous use of household products containing AEs, contributes to NPE and NP loads in urban sewer waters [13,15].

Certain AEs are now used as an alternative to alkyl polyphenol ethoxylates (APEOs), as they are safer as result of their rapid breakdown to less toxic, less persistent and less estrogenic compounds. Generally, the presence of nonionic surfactants in irrigation water leads to two major concerns: (1) The degradation products of non-ionic surfactants such as nonylohenol ethoxylates (NPEO) and NP act as estrogen mimicking compounds, leading to symptoms like change in sex ratio, decrease in fish population and malformations in some aquatic animals [16]. In terms of the estrogenic effects, a study in US raised a great concern of potable water contamination by NPs which have been detected in some wells. The authors reported that the source of these compounds is the continuous infiltration disposal of wastewater treatment plant effluents [17]. (2) Non-ionic surfactants in the irrigation water may have an impact on the mobility of chemicals in soil and cause contamination of groundwater [18,19,20,21]. Katagi (2008) [22] reported that the downward movement of pesticides in soil was enhanced by the presence of high non-ionic surfactant concentrations in leaching water. However, the effect may depend on the surfactant concentration. Similarly, Tao *et al.* (2006) [23] indicated that, at low concentrations 0–20 mg L^{-1}, surfactants have the potential to decrease sorption and thus mobilize herbicides such as atrazine. For example, at low concentrations (0.04 g kg^{-1}), non-ionic surfactants decreased the mobility of the herbicide metolachlor in soil; however, at high concentrations (5–50 g kg^{-1}), the mobility increased [24]. A lysimeter study revealed that at a low concentration (12 mg L^{-1}), NP had no effect on the mobility of atrazine, metolachlor and metribuzin in sandy soil [25].

Pesticide contamination of water bodies has generated a great deal of concern. Generally, it results from diffuse (non-point) sources such as runoff and leaching from agricultural lands. Triazinones herbicides, such as metribuzin (4-amino-6-*tert*-butyl-4,5-dihydro-3-methyltio-1,2,4-triazin-5-one), are used worldwide to control broadleaf weeds in crops such as potatoes (*Solanum tuberosum* L.), soybeans [*Glycine max* (L.) Merr.] and other vegetable crops. Highly water soluble (1050 mg L^{-1}) and having a low adsorption coefficient (K_{oc} = 60), metribuzin shows moderate persistence in soil. Consequently, it may have the potential to leach into the soil profile and thus contaminate groundwater. Further, metribuzin was detected at the highest concentration of 25 µg L^{-1} in Lack Erie tributaries [26] and found to be the most frequently detected herbicide with a maximum concentration of 0.351 µg L^{-1} in 20 samples from drinking water wells in Brazil [27]. The half-life of metribuzin in soils ranges from 11 to 46 days in lab and field experiments; however, it was reported to be as long as 60–90 days under winter conditions in Manitoba [28]. In terms of the toxicity to aquatic plants, Fairchild *et al.* (1998) [29] stated that compared to atrazine, metolachlor and alachlor, metribuzin exhibits the highest level of toxicity. Similarly, metribuzin is highly toxic to freshwater macrophytes and algae under laboratory conditions (at EC_{50} = 31 µg L^{-1}) [30]. Recently, Plhalova *et al.* (2012) [31] investigated the effect of metribuzin in surface water on fish and found that the lowest observed effect concentration (LOEC) is 33 mg L^{-1}.

Studies have shown that metribuzin dissipation is primarily dependent upon degradation by soil microorganisms [32,33,34]. Non-ionic surfactants are commonly used in herbicide formulations to enhance the herbicide penetration through the plant cuticle and thus increase its efficiency by reducing the loss of applied herbicide [35]. Since irrigation with wastewater can introduce large amounts of surfactants to soil, this could enhance the mobility of pesticides, such as metribuzin in soil water systems, by increasing their water solubility. To the best of the authors' knowledge, no study has investigated the fate and transport of metribuzin in the presence of rich surfactant irrigation water under field conditions and the potential contamination of ground water. Therefore, the aim of this study was to investigate the effect of the presence of the non-ionic surfactant Brij35 in irrigation wastewater at concentration above CMC on the downward movement of metribuzin in the soil profile and leachate.

MATERIAL AND METHODS

Chemicals and Reagents

A pure standard (Pestanal®) and formulation (Sencor 75) of metribuzin; trifluralin, used as an internal standard; the non-ionic surfactant Brij 35 [Polyoxyethylene lauryl ether $(C_2H_4O)_{23}$–$C_{12}H_{25}OH$] with critical micelle concentration (CMC) = 74 mg L^{-1}; and HPLC-grade methanol and acetonitrile were obtained from Sigma-Aldrich. Mobile phase chemicals were purchased from Fisher Scientific. Double–deionized water (Milli-Q, Millipore, Molsheim, France) was used in the preparation of standard solutions and mobile phase solutions.

Soil Characteristics

Belonging to the St. Amble complex, the sandy soil used was obtained from a field in the Macdonald Campus of McGill University, Ste-Anne-De-Bellevue, Quebec. Physical and chemical properties of soil are given in Table 1

Table 1: Physical and chemical characteristics of soil

Soil Type	Sand (%)	Silt (%)	pH	Bulk Density (Mg m^{-3})	Organic Matter (%)	CEC (cmol kg^{-1})	Hydraulic Conductivity (cm Day^{-1})
Sandy	92.2	4.3	5.5	1.350	2.97	4.9	1.67 (* SD = 0.45)

* SD denotes standard deviation

Sorption Study

Sorption of metribuzin onto soil in the presence of the non-ionic surfactant Brij 35 was investigated through a soil equilibrium technique. Triplicate soil samples (1 g) were equilibrated with an aqueous solution (10 mL) of metribuzin at concentrations of (2.5, 5.0, 7.5 and 12.5 μg mL^{-1}) in factorial combination with Brij 35 concentrations of 0, 0.5 and 5 g L^{-1}. All solutions contained 0.01 M $CaCl_2$ to mimic the ionic strength of groundwater [36]. The soil slurry was placed in a 50 mL centrifuge tube, and placed on a reciprocal shaker, operating at 200 rpm, for 24 h. The slurry was centrifuged at 2683 g for 20 min. Extraction and analysis of metribuzin proceeded, as described above. The proportion of metribuzin adsorbed was calculated by taking the difference between the amount initially present in solution and the amount remaining in solution, after equilibrium with the soil.

The sorption coefficient was calculated as follows:

$$Kd = \frac{Cs}{Caq} \quad (1)$$

where C_s is the amount of metribuzin sorbed (μg g^{-1}), C_{aq} is the solution concentration in μg mL^{-1}, and K_d is the sorption coefficient (mL g^{-1}).

Organic carbon normalized coefficient values for metribuzin in the absence of surfactant and in the presence of surfactant were obtained as follows:

$$K_{oc} = \frac{K_d}{f_{oc}} \quad (2)$$

where f_{oc} is the organic carbon fraction of the soil (0.0297 g g^{-1} for the soil used), and K_{oc} is the partition coefficient of the contaminant in the organic fraction of the soil (mL g^{-1}).

EXPERIMENTAL SET UP

The field experiment investigating metribuzin transport in a sandy agricultural soil was conducted in nine outdoor PVC lysimeters set up at the Macdonald Campus of McGill University, Ste-Anne-De-Bellevue, Quebec. The lysimeters were irrigated with three different levels of surfactants and applied in each lysimeter using the completely random design. The lysimeters were (0.45 m O.D. × 1 m long), sealed at the bottom to 0.6 m × 0.6 m PVC sheets. Each was packed in layers with a sandy soil and adjusted to a bulk density of 1.35 mg m^{-3}. A 0.05 m diameter drainage pipe was installed at the bottom of each lysimeter. Four, 10 mm diameter, soil sampling holes were made in each lysimeter at depths of 0.1, 0.3 and 0.6 m the soil surface (Figure 1). The lysimeters were sheltered to prevent entry of natural precipitation.

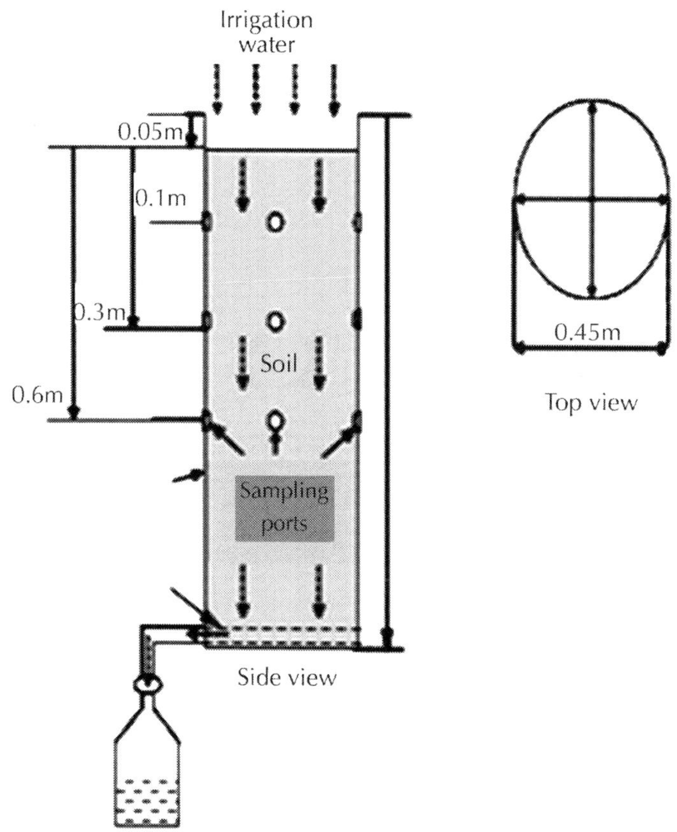

Figure 1: Schematic design of the lysimeter.

APPLICATION OF TESTED COMPOUNDS

Cattle manure was collected from an organic farm in Ham-Sud, Québec, at the end of June 2009. Before the start of the experiment, all lysimeters were irrigated, bringing them to saturation. The drainage pipe at the bottom of each lysimeter was left open throughout the study. The next day (Day 0), with the lysimeter's soil at field capacity, homogenized manure (10 mg ha^{-1}) was applied to the surface of all lysimeters and manually mixed into the top 30 mm of soil [37]. Metribuzin was then sprayed on to each lysimeter's soil surface (1 July 2009) at the locally recommended rate for potatoes (1 kg a.i. ha^{-1}). Twelve liters (70 mm) of irrigation water, containing one of three different concentrations of Brij35 (0, 0.5 or 5.0 g L^{-1}) was applied on day 0, day 21, day 42, day 63, according to recommended irrigation practices for potatoes. Each treatment was replicated three times. While these surfactant concentrations may appear to be high, they are well within concentrations found in wastewater and industrial wastewater. Indeed, concentrations of up to 10 g L^{-1} have been used to study the effect of surfactants on atrazine movement in soils [38].

Soil and Leachate Sampling

Soil samples were collected at the surface as well as 0.1, 0.3 and 0.6 m beneath it, through 4 sampling ports in the side of the lysimeters. Soil samples were collected on 8 occasions: 0, 1, 5, 11, 22, 43, 60 and 90 days after the application of manure. For each surfactant concentration, three replicate soil samples (3 lysimeters) were taken at each of the four depths, and samples from a common treatment × depth were combined in a composite sample (~15 g). Subsamples of about 5 g were taken from each composite sample, and their moisture content was determined. The remaining portions of the samples were stored in sealed bags in the freezer at −4 °C until extraction. Leachate samples (1 L) were collected as a subsample of a total of approximately 9–11 L of leachate, collected at the outlet at the bottom of each lysimeter, following each of the four irrigation events (Figure 1). The subsamples were then transported to the lab and immediately extracted to prevent any chance of metribuzin degradation.

Mass Balance Calculations

At each sampling date, total herbicide mass, recovered from each lysimeter, was calculated as the sum of herbicide recovered in soil samples across all depths of the soil profile. Adding the herbicide mass in the leachate to that in the soil would total the initial herbicide mass applied, minus any losses (unrecovered metribuzin) due to degradation or volatilization (Equation (3)) [39].

$$MTR_{init} = 1590.4\rho\left[C_{(0-5)}\theta_{(0-5)}h_{(0-5)} + C_{(5-15)} + C_{(15-45)}\theta_{(15-45)}h_{(15-45)} + C_{(45-70)}\theta_{(45-70)}h_{(45-70)}\right] + [C_{leach}V_{leach}] + MTR_{lost} \quad (3)$$

where, (3) MTR_{init}: Total metribuzin (mg) initially applied; MTR_{lost}: Metribuzin lost through degradation, volatilization, *etc.* (mg); ρ: Soil bulk density, (g cm^{-3}); $C_{(x-y)}\,\theta_{(x-y)}h_{(x-y)}$: Metribuzin concentration and soil moisture content, respectively, in soil layer h ranging from depth x to depth y (mg g^{-1}); C_{leach}: Metribuzin concentration in leachate (mg L^{-1}); V_{leach}: Volume of leachate; 1590.4: Soil layer surface area, $\pi\left(\frac{D}{2}\right)^2$, where D = 45 cm.

In order to calculate the mass of metribuzin recovered in the soil layer samples, the following equation was used for this purpose:

$$\text{Mass of MTR} = [C\,(m)] \times \rho \times a \times h \quad (4)$$

where C, laboratory reported analytical metribuzin concentration in soil samples (mg g^{-1}); m, moisture content of soil samples (mass water/mass dry soil mg g^{-1}); ρ, soil bulk density (g cm^{-3}); a, area of lysimeter (m^2) and h, sampling depth of soil layers (cm).

The half-life $t_{½}$ of metribuzin was calculated as follows:

$$t_{\frac{1}{2}} = \frac{In2}{K} \quad (5)$$

Then Equation (6) is used in order to calculate k:

$$C = C_0 \, e^{-kt} \qquad (6)$$

where C, the metribuzin concentration at time (t); C_0, the initial metribuzin concentration and k is the metribuzin degradation rate constant.

ANALYTICAL METHODS

Leachate Sample Extraction

Each 1 L subsample was filtered through 90 mm filter paper, followed by filtration through a 45 mm filter (Advantec, Dublin, CA, USA), to remove coarse and fine suspended matter, prior to extraction. Metribuzin-binding MCX cartridges (Oasis Co., Ltd., Milford, MA, USA) were used in solid phase extraction (SPE). Prior to this extraction, trifluralin was added to each of the filtered samples as an internal standard. Preconditioning of each cartridge began by passing 5 mL of water through each cartridge twice, then 5 mL of 50:50 water:methanol (*v/v*), followed by 100% methanol, all at a flow rate of 10 mL min^{-1}. The sample filtrate was then passed through the cartridges then washed twice with 5 mL of water, to elute the metribuzin and trifluralin. Collected in a test tube, the 10 mL of eluate was evaporated under a N_2 stream, residue redissolved in 1 mL of acetonitrile, then passed through 0.22 µm syringe driven filter (Millex-GV, Billerica, MA, USA) ,and filtered solution transferred to a vial for HPLC analysis.

Soil Sample Extraction

In a 50 mL centrifuge tube, to each soil sample (10 g FW), 20 mL of ACS methanol was added, along with trifluralin as an internal standard, and the slurry was votexed for 30 s. The tube was then placed in an ultrasonic bath for 15 min, and then centrifuged at 2683 g for 20 min. The supernatant was decanted and the extraction procedure was repeated twice more. The combined supernatants were passed through a 45 mm filter paper, and then evaporated under a N_2 stream. Samples were redissolved in 1 mL of acetonitrile and passed through 0.22 μm syringe driven filter prior to HPLC analysis.

Analysis

The soil samples and leachate were analyzed by HPLC, equipped with diode array detection (DAD) and an Eclipse plus C_{18} column (4.6 × 150 mm; Agilent, Santa Clara, CA, USA). The mobile phase was acetonitrile and water, run as a gradient; 50:50 acetonitrile:water (*v/v*) for 9 min then change to 10:90 acetonitrile:water (*v/v*) for 8 min. The flow rate was 1 mL min^{-1} and the detection wavelengths were 238 nm and 254 nm for metribuzin and trifluralin, respectively. The extraction efficiency found to be 93% (representing 3 replicates from each concentration) and the detection limit of this method for the tested compound (metribuzin) was 10 ng g^{-1}.

Data Analysis

A statistical model using repeated measures over time and depth was employed to determine if the metribuzin concentrations differed between treatments, over time and over depth. Data was analysed with PROC MIXED in SAS v. 9.2 [40].

RESULTS AND DISCUSSION

Sorption

Among many factors affecting the transport of chemicals in soil, sorption plays an important role; therefore, sorption experiments in the presence and absence of Brij35 were carried out. Results revealed that the amount of metribuzin sorbed was higher in the absence of Brij35; however, it decreased in the presence of Brij35 at concentrations above CMC. The sorption isotherms of metribuzin in tap water and Brij35 solutions at concentrations of 0.5 and 5 g L^{-1} (Figure 2) all showed a good fit ($R^2 \geq 0.97$) with Equation (1). The K_d value for metribuzin shows the highest 1.80 in the absence of the surfactant; however, it decreases significantly to (0.93 and 1.38) in the presence of Brij35 at 0.5 and 5 g L^{-1}, respectively (Table 2). This indicates that the presence of micelles may enhance the release of sorbed metribuzin into the solution. Furthermore, a study by Sun *et al.* (1995) [41] revealed that non-ionic surfactants at concentration above CMC, decreased the sorption coefficient K_d of hydrophobic organic compounds (HOCs). Similarly, Chappell *et al.* (2005) [42] indicated that Brij35 at concentration above CMC competes with atrazine for sorption sites in soil leading to a significant decrease in atrazine sorption. Furthermore, the presence of surfactants enhanced the solubility of Trichloroethylene (TEC) and therefore, higher desorption rate of TEC in the aqueous medium has been observed [43].

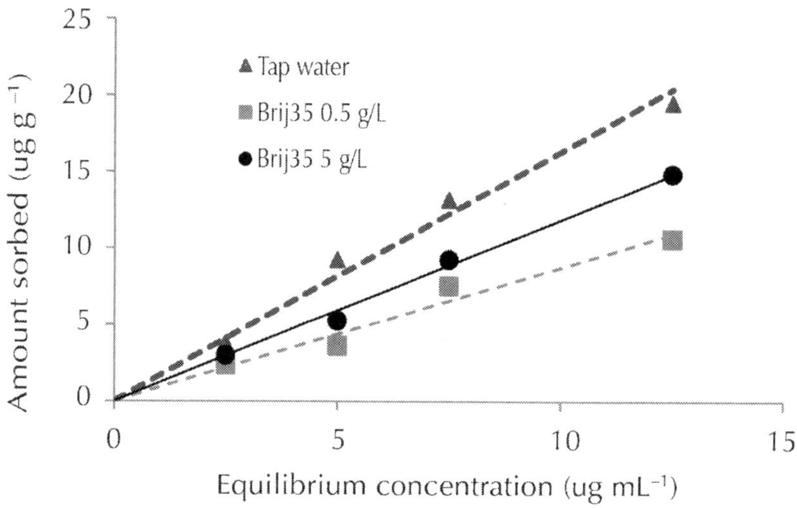

Figure 2: Metribuzin isotherms in the absence and presence of the non-ionic surfactant Brij35.

Table 2: Distribution coefficient Kd, R2 and Koc values for metribuzin sorption in the presence of the nonionic surfactant Brij35

Concentration of Brij35 (g L^{-1})	K_d	R^2	K_{oc}
0	1.8	0.97	60.61
0.5	0.93	0.95	31.31
5	1.38	0.99	46.46

As a result of increasing the concentration of Brij35 from 0.5 to 5 g L^{-1}, K_d values increased from 0.93 to 1.38. This slight increase could be explained by the fact that at higher concentrations, nonionic surfactant sorbed strongly to the soil and formed bilayers and since metribuzin has an affinity to partition with the hydrophilic head of Brij35, presumably via hydrogen bonding. This may lead to enhance metribuzin partitioning into the surfactant which is already sorbed to soil [44,45]. Therefore, increasing the K_d values at the highest Brij35 concentration of 5 g L^{-1} was observed. Furthermore, Nilufar (2005) [25] stated that the sorption coefficient of metribuzin (K_d) was 41% lower in the presence of the nonionic surfactant degradation product nonylephenol (NP) below CMC, confirming the desorption potential of NP for metribuzin.

The aqueous concentration of metribuzin increased once CMC is exceeded. This enhances micelles to form in the solution and compete with the sorbed surfactant for metribuzin. The more micelles present in the aqueous phase, the more metribuzin partitioning into the micelles within the solution, resulting in greater amount released from sorbed metribuzin. The present results are in agreement with the findings of [41,46]. Also, our results are in good match with Genari *et al.* (2009) [47] who indicated that the interaction between the surfactant and soil led to a competition with pesticides for adsorption sites, therefore a decline in pesticides adsorption was observed.

The distribution coefficient K_{oc}, normalized to the corresponding soil organic carbon fraction f_{oc} and calculated from Equation (2), is found to follow the same trend as K_d values. Fundamentally, K_{oc} values have been used as an indicator of the leaching potential of herbicides such as atrazine and metribuzin [48]. In fact the K_{oc} value of 60.61 for metribuzin in the absence of Brij35 has been documented in the literature [49,50,51]. Obviously, the calculated K_{oc} values show a higher leaching potential for metribuzin in the presence of Brij35 (31.31–45.46) compared to the value of 60.61 in the absence of Brij35. This highlights the potential of non-ionic surfactants to enhance the desorption of metribuzin, either by solubilization or partitioning into the surfactant's micellar core.

MASS BALANCE

The mass balance for metribuzin is presented in Table 2. For all irrigation treatments, metribuzin concentrations were found to be below the detection limit by day 90. In both the water and surfactant-irrigated treatments, the total amount of metribuzin detectable in the system (soil + leachate), declined from an initial (applied) level of about 20 mg to less than 3 mg after 63 days and four irrigations. The relative amount of metribuzin in the two lower soil depth layers (15–45 and 45–70 cm) to that in the upper layers (0–5 and 5–15 cm) was at least 2-fold greater in the presence of surfactants from Day 5 onwards, suggesting that a greater quantity of metribuzin leached downwards in the presence (*vs.* absence) of the surfactant. This will be further addressed in the discussion of metribuzin residue concentrations in the soil which follows.

Table 3: Amounts of metribuzin (mg) in different soil profile depth ranges, cumulated in leachate, and overall, over a 63 day period. Soil samplings on days 0, 1, 15, 22, 30, and 60 at depths of 2, 10, 30 and 60 cm depths, as well as and irrigations followed by leachate sampling on Days 1, 21, 42, 63. (—, metribuzin was below detection levels)

Day	Irrigation Solution Soil Profile Depth Range (cm)/Leachate/Total																	
	Tap water						Brij35 at 0.5 g L^{-1}						Brij35 at 5 g L^{-1}					
	0–5	5–15	15–45	45–70	Leach cumul	Total	0–5	5–15	15–45	45–70	Leach cumul	Total	0-5	5-15	15-45	45-70	Leach cumul	Total
0	19.25	—Irrigation I—			0	19.25	19.85	—Irrigation I—			0	19.85	20.06	—Irrigation I—			0	20.06
1	9.62	8.90	—	—	0.09	18.61	10.30	2.13	5.72	—	0.34	18.49	8.49	2.75	6.65	—	0.65	18.54
5	5.05	3.08	0.94	—	0.09	9.16	4.91	1.79	2.55	—	0.34	9.59	2.82	2.26	3.52	—	0.65	9.25
15	2.59	1.05	1.95	—	0.09	5.68	1.88	0.11	4.93	1.99	0.34	9.25	1.54	0.33	2.44	3.29	0.65	8.25
21	——Irrigation II——				0.17	5.76	——Irrigation II——				0.80	9.71	——Irrigation II——				1.27	8.87
22	1.32	0.21	1.36	—	0.17	3.06	0.78	—	1.60	1.72	0.80	4.90	0.95	—	0.91	1.09	1.27	4.22
30	0.85	—	—	—	0.17	1.02	0.62	—	0.82	0.56	0.80	2.80	0.22	—	1.20	1.08	1.27	3.77
42	——Irrigation III——				0.25	1.10	——Irrigation III——				1.33	3.33	——Irrigation III——				1.55	4.05
60	0.12	—	—	—	0.25	0.37	0.09	—	—	0.93	1.33	2.35	0.07	—	—	0.17	1.55	1.79
63	——Irrigation IV——				0.25	0.37	——Irrigation IV——				1.54	2.56	——Irrigation IV——				1.55	1.79

From 15 days onward, the total amount of measured metribuzin was higher in the treatments of surfactant compared to the tap water (Table 3). After 60 days, the total residual metribuzin was 4 and 6 fold greater in the 5 and 0.5 g L^{-1} treatments, respectively, than in the tap water treatment. The percentage of unrecovered metribuzin (Figure 3) was similar for all three irrigation types up to 5 days; however, later values were smaller in the presence of surfactant. As a result, greater recoveries have been measured in the presence of the non-ionic surfactant Brij35 compared to tap water treatment. Despite the fact that metribuzin is mobile in soil, previous studies detected no amount of metribuzin in leachate from lysimeters packed with sandy soil and irrigated with tap water [25,52]. However, in the present study, 0.34% of the applied metribuzin was recovered in the leachate of lysimeters irrigated with tap water, and 1.86% and 2.04%, respectively, for the 0.5 and 5.0 g L^{-1} surfactant treatments.

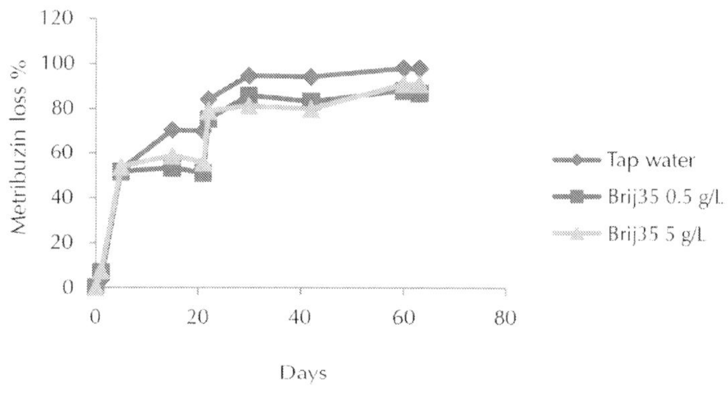

Figure 3: Metribuzin lost percentage in the season in the absence and presence of surfactant.

Degradation of pesticides has been extensively studied and revealed the fact that some factors other than leaching, can significantly affect the breakdown of metribuzin in soil and water. These factors could be sorption, microbial degradation and photolysis [53]. In general, several studies have indicated that the microbial activity plays a major role in the degradation process of metribuzin in soil [54,55]. More specifically, soil microorganisms have the ability to transform metribuzin to its metabolites; desaminometribuzin (DA), desaminodiketometribuzin

(DADK), and diketometribuzin (DK) [56]. In sewage effluents, In the case of non-ionic surfactants, they are subject to biodegradation and produce intermediates which are generally less water soluble such as polyethylene glycols (PEG) and carboxylic aerobic metabolites *i.e.*, carboxylated PEGs [57].

In terms of the amount of unrecovered metribuzin in soil, results showed that in the presence of surfactant, the amount of unrecovered metribuzin was less than the one in the tap water treatment indicating that the non-ionic surfactant could have influenced metribuzin degradation in soil. As indicated in (Figure 3) the presence of surfactant reduced the loss of metribuzin. This may be explained by the findings of Laha and Luthy (1992) [58] that showed a decrease in phenanthrene degradation in the presence of non-ionic surfactants above CMC compared to those below CMC. They assumed that micelles can hold the pesticide from transporting to bacterial cells resulting in an inhibition of bactaerial mineralization of the pesticide. Similarly, comparing to Brij30 and Triton X-100, Brij35 reduced the bioavailability of phenanthrene as a result of the deficient interaction between the hydrophilic moiety of Brij35 (long polyethoxy chain) and bacterial cell membrane [59]. This assumption has been confirmed by Allen *et al.* (1999) [60] who found that cell disruption of microbes is the main cause of inhibition of PAHs degradation, in the presence of nonionic surfactant, Triton X-100.

Effect of Nonionic Surfactant Brij35 on Metribuzin Residues in Soil

The repeated measures analysis showed the effect of treatment (surfactant concentration), depth and time, as well as their single and multiple interactions were all significant ($p \leq 0.05$). Mean metribuzin concentrations in soil over a 90-day period for the soil surface and at 0.1, 0.3 and 0.6 m soil depths are illustrated in Figure 4, Figure 5, Figure 6, Figure 7. As expected, the metribuzin concentration was highest at the soil surface and dropped significantly as sampling depth increased. At the soil surface, the concentration decreased as the degradation rate increased with time. The concentrations right after the first application on day 1 were 968.68, 1036.34 and 854.93 μg kg–1 in the treatments with tap water, Brij35 0.5 g L–1 and Brij35 5 g L–1, respectively. Metribuzin levels declined over time and reached 12.37 μg kg–1 in the topsoil on day 60 for the tap water treatment, whereas for Brij35 0.5 g L−1 and Brij35 5 g L−1, concentrations were lower at 9.82 and 7.29 μg kg−1, respectively. By day 90, metribuzin concentration in the topsoil was below the detection limit for all irrigation treatments. However, a considerable amount of metribuzin leached to the next depths in the treatments with surfactants. This result for the tap water treatment concurs with the findings of several studies [61,62,63,64,65]. They reported that metribuzin concentrations declined in topsoil immediately after application. A number of studies suggest that a significant portion of this decline can be attributed to the presence of soil microorganisms, which play a significant role in the breakdown of metribuzin in soil [66,67]. For all treatments, metribuzin degradation is shown to follow the first order kinetic [68,69]. Typically, a rapid initial dissipation is followed by a gradual decline in the degradation rate and eventually a relatively slow long term breakdown. The half-life of metribuzin under the experiment conditions is calculated. In the tap water treatment, the half- life of metribuzin was 8.89 days; however, it decreased to 8.35 and 7.96 days in the treatment of surfactant at 0.5 and 5.0 g L−1, respectively. This is consistent with the reported half -life by Webster and Reimer (1979) [28].

At the 0.1 m depth, initial metribuzin concentrations in the tap water treatment were higher than those in either of the Brij35 treatments. For all treatments, metribuzin levels decreased over time; however, in the tap water treatment, metribuzin was detectable through day 22, whereas in the Brij35 treatments it only remained detectable through day 15. This disappearance of metribuzin could be attributed to its movement to lower depths and to the microbial degradation. Metribuzin breakdown varies according to soil layers [69] due to the differences in their microbial populations. Indeed, Bouchard et al. (1982) and Danial et al. (2002) [62,70] observed that metribuzin degradation is decreased by depth. Likewise, Fan (2010) [52] conducted a lysimeter study and observed that a considerable amount of surface-applied metribuzin had migrated through a sandy soil to a depth greater than 0.1 m within the first 7 days, and then decreased gradually afterwards. Similarly, results from a study investigating the mobility of metribuzin under free drainage conditions, revealed that a significant portion of surface-applied metribuzin was detected below a 0.1 m soil depth, six days after application [71].

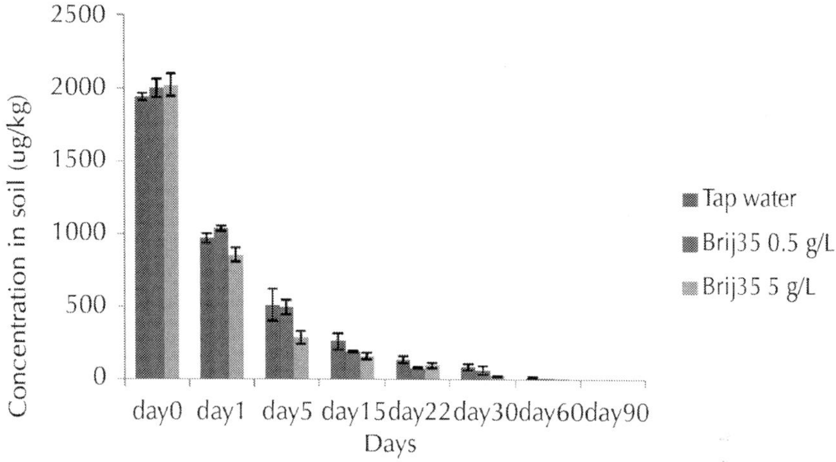

Figure 4: Metribuzin concentrations at the surface.

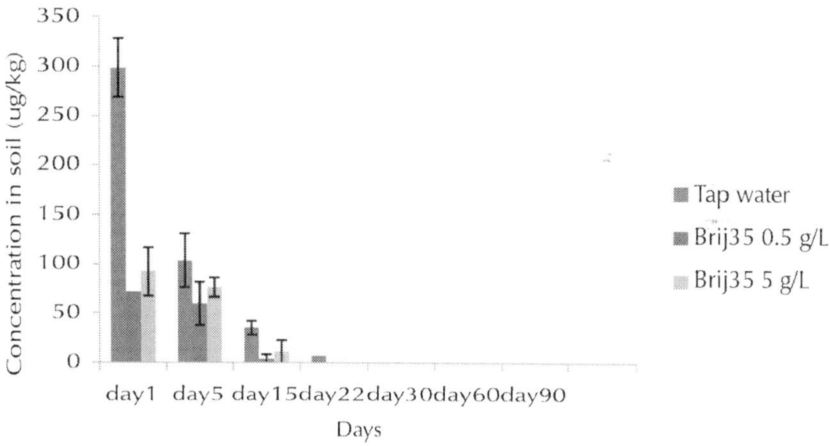

Figure 5: Metribuzin concentrations at 0.1 m depth.

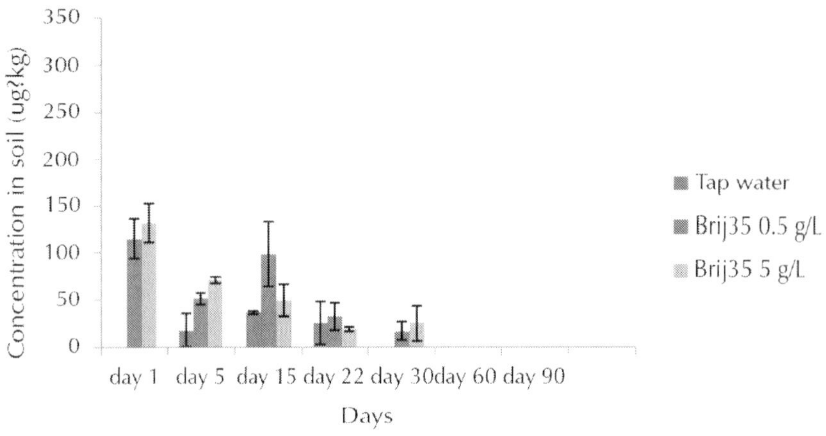

Figure 6: Metribuzin concentrations at 0.3 m depth.

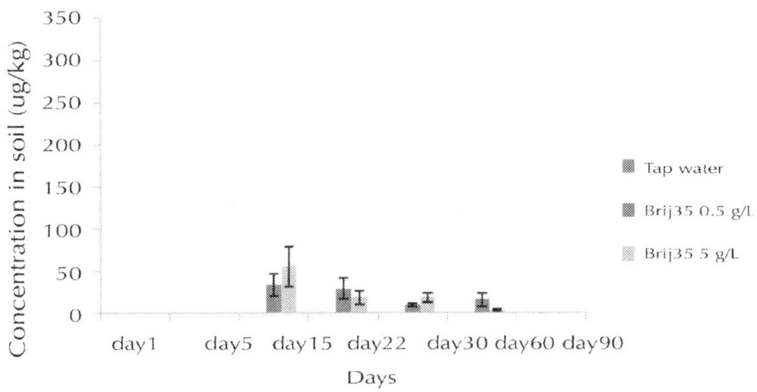

Figure 7: Metribuzin concentrations at 0.6 m depth.

At the 0.3 m depth, the effect of surfactant treatments contrasted with the 0.1 m depth, since metribuzin concentrations were higher in the presence of surfactant compared to the tap water (Figure 6). This indicates that Brij35 influenced the downward movement of metribuzin. Given that metribuzin is a non-ionic compound, belonging to the weakly basic triazines, its sorption to soil is dependent on the amount of organic matter, clay content and pH. It was reported that metribuzin showed a great potential for leaching in soil [51,70]. At the 0.6 m depth, low concentrations of metribuzin (33.46–2.87 μg kg−1) were detected from day 15 to day 60, in the Brij35 treatments, but remained below detection limits for the tap water treatment. This highlights the role of surfactants in enhancing metribuzin mobility in the soil profile (Figure 7). The decrease in metribuzin concentrations over time and depth was reflected by the significant effects ($p \leq 0.05$) of treatment, depth, time and their interactions. Results for tap water concurred with the results of columns studies conducted by Bedmar *et al.* (2004) and Maqueda *et al.* (2009) [51,67]. They indicated the absence of metribuzin in soil samples from soil depths exceeding 0.30 m. Moreover, as the solubility of the herbicide increased, its ability to compete with organic surfaces decreased. This accounts for why, in certain conditions, the migration of metribuzin downward through the soil profile was improved [72].

Generally, surfactants play an important role in enhancing the mobility of pesticides and this fact has been confirmed by Muller *et al.* (2007) and Zhang *et al.* (2011) [43,73]. Our study also showed a greater downward movement of metribuzin through a sandy soil in the presence (*vs.* absence) of the non-ionic surfactant Brij35. Generally, surfactants can increase the solubility of pesticides through their solubilization in the micelles, which formed above the critical micelle concentration (CMC) of the surfactant. Depending on this concentration, pesticide properties, the degree and the location of solubilization into the micelle, surfactant effects can vary. Since metribuzin is a polar herbicide, it is expected to be solubilized by the surfactant once CMC is exceeded. Furthermore, depending on their hydrophilicity, non-ionic surfactants have the ability to solubilize polar compounds [74]. This hydrophilicity is known to be governed by the number of the ethylene oxide groups (EO) in the surfactant molecule, *i.e.*, the greater the number of EO groups, the more hydrophilic the surfactant [75]. Since Brij35 has twenty-three EO groups, it is expected to exhibit a high capacity to solubilise polar compounds, such as metribuzin in the environment.

Effect of Non-Ionic Surfactant Brij35 on Metribuzin in Leachate

The leachate samples were analyzed to quantify the amount of metribuzin leaching to shallow ground water. Metribuzin concentrations in the leachate of both Brij35 treatments, ranging from 67.22 to 23.02 for Brij35 at 0.5 g L^{-1}, and 84.80 µg L^{-1} to below the detection limit for Brij35 at 5 g L^{-1} (Figure 8). The levels of metribuzin in the leachate were significantly greater at the treatment of Brij35 at 5.0 g L−1 than Brij35 at 0.5 g L−1 after the first two irrigation events. Metribuzin concentrations reached a maximum at the third application (17.19 µg L−1) in the tap water leachate samples. This observation can be explained by the fact that metribuzin adsorption in sandy soil is reversible and thus residues might escape to lower layers in the soil profile [67]. Likewise, since sandy soil possess a low CEC and large pore spaces, high leaching potential for herbicides may be observed [76]. However, in a lysimeter study, no metribuzin concentrations were detected in the leachate from a sandy soil after irrigation with either tap water or a non-ionic surfactant metabolite, NP (12 mg L^{-1}) [25].

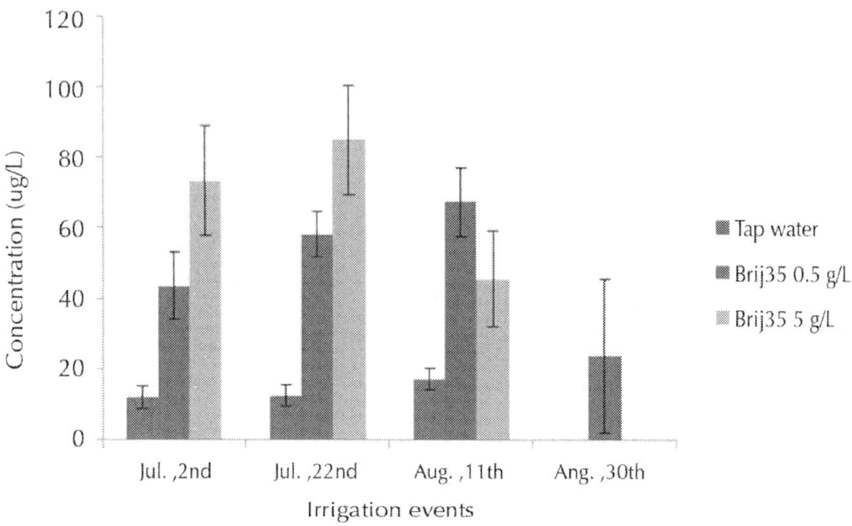

Figure 8: Metribuzin concentrations in Leachate.

In the case of surfactant treatments, the amount of metribuzin moved from the surface to the leachate in the treatment of Brij35 at 5 g L^{-1} was 84.80 µg L^{-1} after the second irrigation event. This amount exceeded the acceptable level for metribuzin in drinking water (80 µg L−1), according to Canadian guidelines [77]. Generally, above the CMC, the lipophilic end (alkyl chain) of the surfactant molecule aggregate together inside the micelle, with the hydrophilic end point towards the aqueous phase on the exterior [78]. Thus, micelles enhance the solubility of low polarity organic compounds by partitioning them into the hydrophobic micelle core. Since metribuzin is a polar herbicide, it is expected to be attracted by the polar head of the surfactant by hydrogen bonding. Partitioning of pesticides with surfactants is well documented in the literature [46,79,80]. More specifically, polar compounds, such as pesticides, are solubilized in the neighborhood of stern layer of the surfactant [22]. Additionally, non-ionic alcohol ethoxylates (AE) are found to enhance the downward movement of pesticides when applied at high (*vs.* low) concentrations to soil columns [18].

Typically, sorption of herbicides onto soil surfaces occurs when the polar components in soil interact with the polar functional groups of the herbicide and since metribuzin has an amide group, interaction with polar sites may occur. However, if water is present, metribuzin tends to interact with water molecules via hydrogen bonding and form hydrophilic complexes, which in turn, facilitate the leaching process. This may explain the relationship between its solubility and leaching potential [81]. Since surfactants can interact with soil, it may compete with metribuzin for available binding sites in soil. This can lead to enhance the free downward movement of metribuzin in soil profile. Recently, Undabeytia *et al.* (2011) [82] confirmed the role of polar groups (amide) in metribuzin in the formation of water bridges via hydrogen bonding with the hydrophilic head of the non-toxic surfactant phosphatidylcholine in clay formulations. Therefore, in the presence of a non-ionic surfactant, such as Brij35at concentrations above its CMC, metribuzin would be captured by micelles, with its polar groups oriented toward the polar head (EO) of the micelles, which would exhibit a strong hydrophilic interaction via hydrogen bonding or dipole-dipole effects. In the present study, results of the lysimeter experiment are explained by the findings of the sorption study. This conclusion confirms the role of non-ionic surfactant Brij35 in decreasing the sorption coefficient of metribuzin and hence promoting its release from soil. Therefore, additional research should be carried out to better understand the effect of chemical compounds when they coexist as mixtures in the environment.

CONCLUSIONS

Understanding the mobility of metribuzin within soil-water-surfactant systems is an important step in evaluating the risk resulting from irrigation with wastewater. The present study showed the presence of non-ionic surfactant (Brij35) in concentrations higher than its CMC, increased the downward movement of metribuzin in the soil profile, presumably due to the solubilization power of micelles in the soil-water system when Brij35 was present at 0.5 or 5 g L^{-1} or the competition between the surfactant and metribuzin for soil binding sites. Additionally, metribuzin concentrations in the leachate were higher (about 5 times) in the presence of Brij35 compared to the tap water. In spite of the fact that metribuzin is water soluble, other lysimeter studies have shown metribuzin to rarely move below 30 cm depth in the soil profile. The present study indicates that the presence of a non-ionic surfactant in irrigation water increased the mobility of metribuzin in soil and led to its presence in leachate samples at a depth of 1 m below the soil. Therefore, the continuous use of wastewater containing non-ionic surfactants at concentrations exceeding their CMC can be expected to increase the possibility of pesticides being detected in leachate, eventually reaching ground water, and thus potentially contaminating water resources.

ACKNOWLEDGMENTS

The authors would like to acknowledge the financial support provided by Faculty of Agriculture, Zagazig University, Egypt and the Natural Sciences and Engineering Research Council of Canada (NSERC). The authors thank R.M. Patel (Research associate in the department of Bioresource Engineering, McGill University, Montreal, Canada) for his guidance in the field experiments.

REFERENCES

1. *Comprehensive Assessment of the Freshwater Resources of the World*; World Meteorological Organization: Geneva, NY, USA, 1997.

2. Rahman, S.H.; Khanam, D.; Adyel, T.M.; Islam, M.S.; Ahsan, M.A.; Akbor, M.A. Assessment of heavy metal contamination of agricultural soil around dhaka export processing zone (DEPZ), bangladesh:Implication of seasonal variation and indices. *Appl. Sci.* 2012, *2*, 584–601.
3. Chatterjee, S.N. Fresh produce from wastewater. *Environ. Sci. Technol.* 2008, *42*, 7732.
4. Drechsel, C.; Raschid-Sally, L.; Redwood, M.; Bahri, A. *Water Irrigation and Health. Assessing and Mitigating Risk in Low-Income Countries*; Earthscan/IDRC: London, UK, 2010.
5. Hamilton, A.J.; Stagnitti, F.; Xiong, X.; Kreidl, S.L.; Benke, K.K.; Maher, P. Wastewater irrigation: The state of play. *Vadose Zone J.* 2007, *6*, 823–840.
6. Brunner, P.H.; Capri, S.; Marcomini, A.; Giger, W. Occurrence and behavior of linear alkylbenzenesulphonates, nonylphenol, nonylphenol mono- and nonylphenol diethoxylates in sewage and sewage sludge treatment. *Water Res.* 1988, *22*, 1465–1472.
7. Field, J.A.; Leenheer, J.A.; Thorn, K.A.; Barber, L.B.; Rostad, C.; Macalady, D.L.; Daniel, S.R. Identification of persistent anionic surfactant derived chemicals in sewage effluent and ground water. *J. Contam. Hydrol.* 1992, *9*, 55–78.
8. Wild, S.B.; Waterrath, K.S.; Jones, K.F. Organic contaminants in an agricultural soil with a known history of sewage sludge amendments. *Environ. Sci. Technol.* 1990, *24*, 1706–1711.
9. Narkis, N.; Ben-David, B. Adsorption of nonionic surfactants on active carbon and mineral clay. *Water Res.* 1985, *19*, 815–824.
10. Castillo, M.; Alonso, M.C.; Riu, J.; Barcelo, D. Identification of polar, ionic, and highly water soluble organic pollutants in untreated industrial wastewaters. *Environ. Sci. Technol.* 1999, *33*, 1300–1306.
11. Feitkenhauer, H.; Meyer, U. On-line titration of non-ionic surfactants in wastewater treatment plants using a specific electrode. *Water Sci. Technol.* 2002, *45*, 61–68.
12. Zoller, U. Non-ionic surfactants in reused water: Are activated sludge/soil aquifer treatments sufficient? *Water Res.* 1994, *28*, 1625–1629.

13. Loyo-Rosales, J.E.; Rice, C.P.; Torrents, A. Fate of octyl and nonylphenol ethoxylates and some carboxylated derivatives in three american wastewater treatment plants. *Environ. Sci. Technol.* 2007, *41*, 6815–6821.

14. Bhandari, A.; Surampalli, R.Y.; Adams, G.D.; Champagne, P.; Ong, S.K.; Tyagi, R.D.; Zhang, T. *Contaminats of Emerging Environmental Concern*; The American Society of Civil Engineers: Reston, VA, USA, 2009.

15. Conn, K.E.; Barber, L.B.; Brown, G.K.; Siegrist, R.L. Occurrence and fate of organic contaminants during onsite wastewater treatment. *Environ. Sci. Technol.* 2006, *40*, 7358–7366.

16. Esperanza, M.; Suidan, M.T.; Nishimura, F.; Wang, Z.; Sorial, G.A. Determination of sex hormones and nonylphenol ethoxylates in the aqueous matrixes of two pilot-scale municipal wastewater treatment plants. *Environ. Sci. Technol.* 2004, *38*, 3028–3035.

17. Barber, L.B.; Keefe, S.H.; Leblanc, D.R.; Bradley, P.M.; Chapelle, F.H.; Meyer, M.T.; Loftin, K.A.; Kolpin, D.W.; Rubio, F. Fate of sulfamethoxazole, 4-Nonylphenol, and 17-Estradiol in groundwater contaminated by wastewater treatment plant effluent. *Environ. Sci. Technol.* 2009, *43*, 4843–4850.

18. Huggenberger, F.H.; Letey, J.; Farmer, W.J. Effect of two nonionic surfactants on adsorption and mobility of selected pesticides in a soil-system. *Soil Sci. Soc. Am. Proc.* 1973, *37*, 215–219.

19. Kan, T.A.; Tomson, M.B. Ground water transport of hydrophobic organic compounds in the presence of dissolved organic matter. *Environ. Toxicol. Chem.* 1990, *9*, 253–263.

20. Aronstein, B.N.; Calvillo, Y.M.; Alexander, M. Effect of surfactants at low concentrations on the desorption and biodegradation of sorbed aromatic compounds in soil. *Environ. Sci. Technol.* 1991, *25*, 1728–1731.

21. Rodriguez-Cruz, M.S.; Sanchez-Martin, M.J.; Sanchez-Camazano, M. Enhanced desorption of herbicides sorbed on soils by addition of Triton X-100. *J. Environ. Qual.* 2004, *33*, 920–929.

22. Katagi, T. Surfactant effects on environmental behavior of pesticides. *Rev. Environ. Contam. Toxicol.* 2008, *194*, 71–177.

23. Tao, Q.; Dong, S.; Hong, X.; Wang, T. Effect of surfactants at low concentrations on the sorption of atrazine by natural sediment. *Water Environ. Res.* 2006, *78*, 653–660.

24. Sanchez-Camazano, M.; Arienzo, M.; Sanchez-Martin, M.J.; Crisanto, T. Effect of different surfactants on the mobility of selected non-ionic pesticides in soil. *Chemosphere* 1995, *31*, 3793–3801.
25. Nilufar, F. Fate and Transport of Herbicides in Soil in the Presence of Surfactants in the Irrigation Water. *Master's Thesis*, McGill University, Montreal, Canada, November 2005.
26. Richards, R.P.; Baker, D.B. Pesticide concentration patterns in agricultural drainage networks in the Lake Erie basin. *Environ. Toxicol. Chem.* 1993, *12*, 13–26.
27. Dores, E.F.G.C.; Navickiene, S.; Cunha, M.L.F.; Carbo, L.; Ribeiro, M.L.; de-Lamonica-Freire, E.M. Multiresidue determination of herbicides in environmental waters from Primavera do Leste region (Middle West of Brazil) by SPE-GC-NPD. *J. Braz. Chem. Soc.* 2006, *17*, 866–873.
28. Webster, G.R.B.; Reimer, G.J. Field degradation of the herbicide metribuzin and its degradation products in a Manitoba sandy loam soil. *Weed Res.* 1976, *16*, 191–196.
29. Fairchild, J.F.; Ruessler, D.S.; Carlson, A.R. Comparative sensitivity of five species of macrophytes and six species of algae to atrazine, metribuzin, alachlor, and metolachlor. *Environ. Toxicol. Chem.* 1998, *17*, 1830–1834.
30. Fairchild, J.F.; Sappington, L.C. Fate and effects of the triazinone herbicide metribuzin in experimental pond mesocosms. *Arch. Environ. Contam. Toxicol.* 2002, *43*, 198–202.
31. Plhalova, L.; Stepanova, S.; Praskova, E.; Chromcova, L.; Zelnickova, L.; Divisova, L.; Skoric, M.; Pistekova, V.; Bedanova, I.; Svobodova, Z. The effects of subchronic exposure to metribuzin on *Danio rerio*. *Sci. World J.* 2012, *2012*.
32. Ladlie, J.S.; Meggitt, W.F.; Penner, D. Effect of soil pH on microbial degradation, adsorption, and mobility of metribuzin. *Weed Sci.* 1976, *24*, 477–481.
33. Sharom, M.S.; Stephenson, G.C. Behaviour and fate of metribuzin in eight Ontario soils. *Weed Sci.* 1976, *24*, 153–160.
34. Khoury, R.; Geahchan, A.; Coste, C.M.; Antoun, M.A. The Behavior of Pesticide in Soil: The Influence of Various Environmental Factors on the Degradation of Metribuzin. In Proceedings of

2000 Mediterranean Conference for Environment and Solar, Beirut, Lebanon, 16–17 November 2000; pp. 34–39.

35. Ramsey, R.J.L.; Stephenson, G.R.; Hall, J.C. A review of the effects of humidity, humectants, and surfactant composition on the absorption and efficacy of highly water—Soluble herbicides. *Pestic. Biochem. Physiol.* 2005, *82*, 162–175.
36. Wilde, D.T.; Mertens, J.; Spanoghe, P.; Ryckeboer, J.; Jaeken, P.; Springael, D. Sorption kinetics and its effects on retention and leaching. *Chemosphere* 2008, *72*, 509–516.
37. Chettri, M.; Thapa, U. Integrated nutrient management with farm yard manure on potatos (*Solanum tuberosum*) under gangetic plains of west Bengal. *Environ. Ecol.* 2004, *22*, 766–769.
38. Abu-Zreig, M.; Rudra, R.P.; Dickinson, W.T.; Evans, L.J. Effect of surfactants on sorption of atrazine by soil. *J. Contam. Hydrol.* 1999, *36*, 249–263.
39. ElSayed, E.M.; Prasher, S.O.; Patel, R.M. Effect of nonionic surfactant Brij35 on the fate and transport of oxytetracycline antibiotic in soil. *J. Environ. Manag.* 2013, *116*, 125–134.
40. *SAS/GRAPH*, 2nd, SAS Institute Inc. Cary, NC, USA, 2010.
41. Sun, S.; Inskeep, W.P.; Boyd, S. Sorption of nonionic compounds in soil-water systems containing a micelle-forming surfactant. *Environ. Sci. Technol.* 1995, *29*, 903–913.
42. Chappell, M.A.; Laird, D.A.; Thompson, M.L.; Evangelou, V.P. Cosorption of atrazine and a lauryl polyoxyethylene oxide nonionic surfactant on smectite. *J. Agric. Food Chem.* 2005, *53*, 10127–10133.
43. Zhang, M.; He, F.; Zhao, D.; Hao, X. Degradation of soil-sorbed trichloroethylene by stabilizedzero valent iron nanoparticles: Effects of sorption, surfactants, and natural organic matter. *Water Res.* 2011, *45*, 2401–2414.
44. Liu, Z.; Edwards, D.; Luthy, R. Sorption of non-ionic surfactants onto soils. *Water Res.* 1992, *26*, 1337–1345.
45. Mata-Sandoval, J.C.; Karns, J.; Torrents, A. Effects of rhamnolipids produced by Pseudomonas aeruginosa UG2 on the solubilization of pesticides. *Environ. Sci. Technol.* 2000, *34*, 4923–4930.

46. Wang, P.; Keller, A.A. Partitioning of hydrophobic organic compounds within soil–water–surfactant systems. *Water Res.* 2008, *42*, 2093–2101.

47. Gennari, M.; Messina, C.; Abbate, C.; Baglieri, A.; Boursier, C. Solubility and adsorption behaviors of chlorpyriphos-methyl in the presence of surfactants. *J. Environ. Sci. Health. Part B* 2009, *44*, 235–240.

48. Southwick, L.M.; Willis, G.H.; Johnson, D.C.; Selim, H.M. Leaching of nitrate, atrazine and metribuzin from sugarcane in southern Louisiana. *J. Environ. Qual.* 1995, *24*, 684–690.

49. Comfort, S.D.; Shea, P.J.; Roeth, F.W. *Understanding Pesticides and Water Quality in Nebraska*; Nebraska Cooperative Extension EC, University of Nebraska-Lincoln: Nebraska, USA, 1994.

50. Weber, J.B.; Keller, K.E. Mobility of Pesticides in Field Lysimeters. In *Mechanisms of Pesticide Movement into Ground Water*; Honeycutt, R.C., Schabacker, D.J., Eds.; Lewis Publishers: Boca Raton, FL, USA, 1994; pp. 43–62.

51. Bedmar, F.; Costa, J.L.; Suero, E.; Gimenez, D. Transport of atrazine and metribuzin in three soils of the humid pampas of Argentina. *Weed Technol.* 2004, *18*, 1–8.

52. Fan, M. Fate and Transport of Herbicides in a Sandy Soil in the Presence of Antibiotics in Poultry Manure. *Master's Thesis*, McGill University, Montreal, Canada, August 2009.

53. Pandiselvi, V.; Sathiyanarayanan, S.; Ayyappan, S.; Ramesh, A. Photolysis of metribuzin in water under direct sunlight—Identification of phototransformation products by LC-MS-MS electrospray tandem mass spectrometry and impact on aquatic species (Pseudokirchneriella subcapitata). *Int. J. Res. Chem. Environ.* 2012, *2*, 251–262.

54. Sabadie, J. Degradation of bensulfuron-methyl on various minerals and humic acids. *Weed Res.* 1997, *37*, 411–418.

55. Jurado-Exposito, M.; Walker, A. Degradation of isoproturon, propyzamide and alachlor in soil with constant and variable incubation conditions. *Weed Res.* 1998, *38*, 309–318.

56. Kjaer, J.; Olsen, P.; Henriksen, T.; Ullum, M. Leaching of metribuzin metabolites and the associated contamination of a sandy danish aquifer. *Environ. Sci. Technol.* 2005, *39*, 8374–8381.

57. Soares, A.; Guieysse, B.; Jefferson, B.; Cartmell, E.; Lester, J.N. Nonylphenol in the environment: A critical review on occurrence, fate, toxicity and treatment in wastewaters. *Environ. Int.* 2008, *34*, 1033–1049.

58. Laha, S.; Luthy, R.G. Effects of nonionic surfactants on the solubilization and mineralization of phenanthrene in soil-water systems. *Biotechnol. Bioeng.* 1992, *40*, 1367–1380.

59. Guha, S.; Jaffe, P.R. Bioavailability of hydrophobic compounds partitioned into the micellar phase of nonionic surfactants. *Environ. Sci. Technol.* 1996, *30*, 1382–1391.

60. Allen, C.C.R.; Boyd, D.R.; Hempenstall, F.; Larkin, M.J.; Sharm, N.D. Contrasting effects of a nonionic surfactant on the biotransformation of polycyclic aromatic hydrocarbons to *cis*-dihydrodiols by soil bacteria. *Appl. Environ. Microbiol.* 1999, *65*, 1335–1339.

61. Savage, K.E. Adsorption and mobility of metribuzin in soil. *Weed Sci.* 1976, *24*, 525–528.

62. Bouchard, D.C.; Lavy, T.L.; Marx, D.B. Fate of metribuzin, metolachlor, and fluometuron in soil. *Weed Sci.* 1982, *30*, 629–632.

63. Fuscaldo, F.; Bedmar, F.; Monterubbianse, G. Perseistance of atrazin, metribuzin and simazine herbicides in soils. *Pesqui. Agropecu. Bras.* 1999, *34*, 2037–2044.

64. Selim, H.M. Retention and runoff losses of atrazine and metribuzin in soil. *J. Environ. Qual.* 2003, *32*, 1058–1071.

65. Henriksen, T.; Svensmark, B.; Juhler, R.K. Degradation and sorption of metribuzin and primary metabolites in a sandy soil. *J. Environ. Qual.* 2004, *33*, 619–627.

66. Villaverde, J.; Kah, M.; Brown, C.D. Adsorption and degradation of four acidic pesticides in soils from southern Spain. *Pest Manag. Sci.* 2008, *64*, 703–710.

67. Maqueda, C.; Villaverde, J.; Sopena, F.; Undabeytia, T.; Morillo, E. Effects of soil characteristics on metribuzin dissipation using clay-gel-based formulations. *J. Agric. Food Chem.* 2009, *57*, 3273–3278.

68. Bowman, B.T. Mobility and dissipation studies of metribuzin, atrazine and their metabolities in plainfield sand using field lysimeters. *Environ. Toxicol. Chem.* 1991, *10*, 573–579.
69. Di, H.J.; Aylmore, L.A.G.; Kookana, R.S. Degradation rates of eight pesticides in surface and subsurface soils under laboratory and field conditions. *Soil Sci.* 1998, *163*, 404–411.
70. Denial, P.E.; Bedmar, F.; Costa, L.J.; Aparicio, V.C. Atrazine and metribuzin sorption in soils of the Argentinean humid pampas. *Environ. Toxicol. Chem.* 2002, *21*, 2567–2572.
71. Jebellie, J.; Prasher, S.O. Role of water table management in reducing metribuzin pollution. *Trans. ASAE* 1998, *41*, 1051–1060.
72. Olness, A.; Basta, N.T.; Rinke, J. Redox effects on resin extraction of herbicides from soil. *Talanta* 2002, *57*, 383–391.
73. Muller, K.; Magesan, G.N.; Bolan, N.S. A critical review of the influence of effluent irrigation on the fate of pesticides in soil. *Agric. Ecosyst. Environ.* 2007, *120*, 93–116.
74. Xiarchos, I.; Doulia, D. Effect of nonionic surfactants on the solubilization of alachlor. *J. Hazard. Mater* 2006, *B136*, 882–888.
75. Jafverti, C.T.; Vanhoof, P.L.; Heath, J. Solubilization of non polar compounds by non ionic surfactant micelles. *Water Res.* 1994, *28*, 1009–1017.
76. Aggarwal, V.; Li, H.; Laird, D.A.; Boyd, S.A.; Johnston, C.T.; Teppen, B.J. Sorption of Triazines and Trichloroethene to Homoionic Smectites. In Proceedings of the 18th World Congress of Soil Science, Philadelphia, PA, USA, 9–15 July 2006.
77. Health Canada. 2010. Available online: http://www.hc-sc.gc.ca/ewh-semt/pubs/water-eau/2010-sum_guide-res_recom/index-eng.php (accessed on 24 August 2011).
78. Cheah, P.S.; Reible, D.; Valsaraj, K.T.; Constant, D.; Walsh, W.; Thibodeaux, L.J. Simulation of soil washing with surfactants. *J. Hazard. Mater.* 1998, *59*, 107–122.
79. Zimmerman, J.B.; Kibbey, T.C.G.; Cowell, M.A.; Hayes, K.F. Partitioning of ethoxylated nonionic surfactants into nonaqueous-phase organic liquids: Influence on solubilization behavior. *Environ. Sci. Technol.* 1999, *33*, 169–176.

ABSTRACT

Nitrogen is a main contaminant of wastewater worldwide. Novel processes for nitrogen removal have been developed over the last several decades. One of these is the partial nitritation process. This process includes the oxidation of ammonium to nitrite without the generation of nitrate. The partial nitritation process has several advantages over traditional nitrification-denitrification processes for nitrogen removal from wastewaters. In addition, partial nitritation is required for anammox elimination of nitrogen from wastewater. Partial nitritation is affected by operational conditions and substances present in the influent, such as quinolone antibiotics. In this review, the impact that several operational conditions, such as temperature, pH, dissolved oxygen concentration, hydraulic retention time and solids retention time, have over the partial nitritation process is covered. The effect of quinolone antibiotics and other emerging contaminants are discussed. Finally, future perspectives for the partial nitritation process are commented upon.

INTRODUCTION

One of the most important environmental problems today is the discharge of wastewater from urban and industrial sources, since human activities have greatly accelerated and extended the natural cycles of nitrogen in the soil, water and atmosphere. In this way, different technologies have been used for the removal of different environmental pollutants, such as nitrogen, organic matter and phosphorus. In wastewater treatment plants, nitrogen is often removed by conventional biological processes, such as nitrification and heterotrophic denitrification. In particular, the conventional processes are not suitable for the treatment of effluents, such as a dewatering concentrate stream, due to their toxic effects to those microorganisms able to degrade the nitrogen. For this reason, alternative systems have been developed for the removal of this nutrient, such as partial nitritation/anaerobic ammonium oxidation (anammox) technology. Thanks to these novel technologies, high ammonium-low carbon effluents, such as reject water from anaerobic digesters, can be treated within the sludge pipeline, therefore avoiding its treatment within the activated sludge processes.

Antimicrobial agents are among the most commonly used pharmaceuticals. However, although antibiotics are of great importance for human health, they can also be a great problem for the environment. During wastewater treatment, antibiotics are drastically removed from the water stream, but their fate is associated with sewage sludge. In this sense, the biological processes of nitritation have been described as particularly sensitive to toxic substances, such as pesticides and antibiotics.

The partial nitritation process is a technique in which nitrification is achieved with nitrite as the intermediate under stable processing conditions, where only 50% of the ammonium in the influent is converted into nitrite. This system and its combination with the anammox process have led to the development of a new technology of great interest in the treatment of effluent with high nitrogen content.

The results obtained in partial nitritation bioreactors show that modifications to the operating conditions can dramatically affect the functioning of the partial nitritation technology. Specifically, a significant factor that produces changes in the microbial communities is hydraulic retention time (HRT) [1]. Thus, changes in the operating conditions, such as HRT, may produce changes in cellular physiology and the community level. Consequently, the performance of the biotransformation of ammonium into nitrite in a partial nitritation bioreactor can be directly affected by the HRT, which can be crucial for the optimization of nutrient removal rates and the implementation of control strategies.

Antibiotics, such as fluoroquinolones, have been reported [2,3] as able to impact the performance of the partial nitritation process, suggesting that there is a negative correlation between antibiotic concentration in the influent and performance of the partial nitritation process. Furthermore, the microbial community inside the bioreactors suffers a period of adaptation or a deep change with a significant reduction of the ammonium-oxidizing bacteria in response to the antibiotic concentration in the influent [3]. Additionally, the presence of antibiotics in the bioreactor increases the number of antibiotic-resistant bacteria. In this way, it could be suggested that the presence of selected antibiotics, such as quinolones, can seriously affect partial nitritation systems and select microorganisms with antibiotic resistance [3].

The objective of our article is to evaluate the influence of different environmental parameters, such as HRT, and the concentration of fluoroquinolone antibiotics on the performance of partial nitritation systems. We also review the effects of these parameters on the microbial communities growing in bioreactors using molecular biology techniques.

Partial Nitritation Process: An Overview

Contamination caused by wastewater discharge has increased during the last decades, due to the increasing global population [1]. Nitrogen, one of the main contaminants of wastewater, has been identified to cause oxygen depletion and eutrophication in aquatic environments [4]. Wastewater engineering has traditionally removed nitrogen from wastewater through complete nitrification-denitrification processing to meet the water quality criteria for treated wastewater [1]. More restrictive standards for nitrogen in wastewater effluent are being imposed by new regulations, such as the European Union (EU) Water Frame Directive 91/271/EEC (Table 1). To achieve these standards in a cost-effective way, autotrophic nitrogen removal technologies, such as partial nitritation/anaerobic ammonium oxidation (anammox), DEMON (deammonification), OLAND (oxygen limited autotrophic nitrification/denitrification), and CANON (completely autotrophic nitrogen-removal over nitrate) [5,6], have been developed. Autotrophic nitrogen removal technologies are based on anammox bacteria, which can eliminate nitrogen by utilizing ammonium as a substrate and nitrite as the terminal electron acceptor in molar ratios of 1:1 [7]. These technologies have been successfully utilized at full-scale wastewater treatment plants for the treatment of reject water from anaerobic digesters (e.g., the partial nitration/anammox process at Rotterdam wastewater treatment plant (WWTP) and the DEMON process at Apeldoorn WWTP and the CANON process at Olburgen WWTP, all located in The Netherlands). Anammox technologies can save up to 90% of costs with respect to traditional nitrification-denitrification processes for influents with a high ammonium concentration and low carbon content [8]. Among these technologies, the partial nitritation/anammox system has been developed as a two-step autotrophic nitrogen removal process. The first step involves the oxidation of ammonium to nitrite under aerobic conditions in such a way that roughly 50% of ammonium is oxidized

[8]. This is achieved with partial nitritation technology. Today, there are six examples of full-scale partial nitritation plants in The Netherlands and one in the USA [9].

Table 1: European Directive 91/271/EEC Requirements in Matter of Nutrients (from [10])

Parameter	Concentration	Minimum percentage of eduction [1]	Reference method of measurement
Total Phosphorus	2 mg/L (10,000–100,000 population equivalent)1 mg/L (more than 100,000 population equivalent)	80	Molecular Absorption Spectrophotometry
Total Nitrogen [2]	15 mg/L (10,000–100,000 population equivalent) [3]10 mg/L (more than 100,000 population equivalent) [3]	70–80	Molecular Absorption Spectrophotometry

Notes: [1] reduction in relation to the load of the influent; [2] total nitrogen means the sum of total Kjeldahl nitrogen (organic and ammoniacal nitrogen), nitrate-nitrogen, and nitrite-nitrogen; [3] these values for concentration are annual means, as referred to in Annex I, paragraph D.4[c]. However, the requirements for nitrogen may be checked using daily averages when it is proven, in accordance with Annex I, paragraph D.1, that the same level of protection is obtained. In this case, the daily average must not exceed 20 mg/L of total nitrogen for all of the samples when the temperature from the effluent in the biological reactor is higher than or equal to 12 °C. The conditions concerning temperature could be replaced by a limitation on the time of operation to account for regional climatic conditions.

In a partial nitritation system, ammonium is oxidized to nitrite (Figure 1). Partial nitritation systems were developed by the Delft University of Technology in 1996–1999 [11] and have several advantages over total nitritation-based technologies, such as 25% savings in aeration, 30% reduction of biomass generation, with a biomass yield of about 0.15 g biomass (g NH_4^+-N)$^{-1}$ [12], and 20% less CO_2 emission [11]. Partial nitritation bioreactors at the laboratory scale have been reported to successfully treat food processing and agriculture industry wastewater, reject wastewater and slaughterhouse wastewater or swine manure wastewater, among others [13,14,15,16,17,18,19].

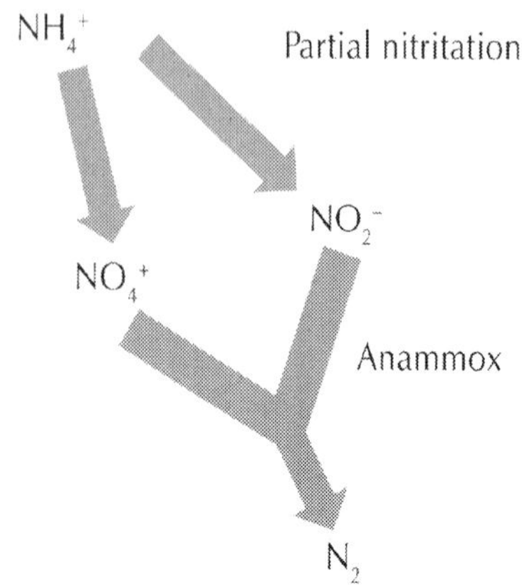

Figure 1: Nitrogen cycle in the partial nitritation/anammox processes.

Within a partial nitritation reactor, the main performance is the oxidation of ammonium to nitrite, and therefore, there is an accumulation of nitrite in the system. Although nitrite has been thought not to accumulate in ecosystems, some reports show that it can accumulate in natural and engineered environments, such as soils, sediments and wastewater treatment plants [20]. This is achieved by the metabolism of ammonium-oxidizing bacteria (AOB).

AOB use ammonium mono-oxidase (AMO) enzyme to oxidize ammonium to hydroxylamine (NH2OH), using oxygen as an electron acceptor. Following this reaction, hydroxylamine is oxidized to nitrite with the mediation of hydroxylamine oxidoreductase (HAO), with hydrazine as an intermediate [11,21]. AOB are autotrophic microorganisms, so they utilize inorganic carbon as a carbon source. AOB communities belong to the β-Proteobacteria class, with species like Nitrosomonas sp., Nitrosospira sp., Nitrosolobus sp. and Nitrosovibrio sp., among others (Table 2). It has also been found that Nitrosomonas and Nitrosospira are the most popular genera among partial nitritation reactors, withNitrosospira dominating under high-ammonium conditions [22]. Two different types of AOB bacteria have been differentiated so far: fast-growing AOB and slow-growing AOB. The difference between these two groups resides in the affinity for ammonium, which is higher in slow-growing AOB. Thus, slow-growing AOB, k-strategists, dominate in environments with an ammonium limitation. In partial nitritation reactors, the ammonium concentration is high, and therefore, fast-growing AOB, r-strategists, dominate [23]. Even though it is known that species of AOB can carry out denitrifying metabolism that reduces nitrite to nitric oxide, nitrous oxide and dinitrogen [9], it is thought that AOB only perform ammonium oxidation in partial nitritation reactors.

Table 2: References for the identification of ammonium-oxidizing bacteria (AOB), nitrite-oxidizing bacteria (NOB) and anammox bacteria in engineered and natural ecosystems

Species	Ecological Role	Class	References
Nitrosomonas sp.	NH_4^+ oxidizer	β-Proteobacteria	[1,11,22,24,25,26]
Nitrosospira sp.	NH_4^+ oxidizer	β-Proteobacteria	[1,22,25]
Nitrosolobus sp.	NH_4^+ oxidizer	β-Proteobacteria	-
Nitrosovibrio sp.	NH_4^+ oxidizer	β-Proteobacteria	[1]
Nitrosococcus sp.	NH_4^+ oxidizer	γ-Proteobacteria	[24,26]
Nitrobacter sp.	NO_2^- oxidizer	α-Proteobacteria	[11,24]
Nitrococcus sp.	NO_2^- oxidizer	γ-Proteobacteria	-
Nitrospina sp.	NO_2^- oxidizer	Nitrospinae	-
Nitrospira sp.	NO_2^- oxidizer	Nitrospirae	[26]
Brocadia sp.	Anammox	Planctomycetia	[27]
Kuenenia sp.	Anammox	Planctomycetia	[28,29]
Scalindua sp.	Anammox	Planctomycetia	[30,31]

Nevertheless, AOB have to compete with other microbial communities inside a partial nitritation reactor. The main competitors of AOB are nitrite-oxidizing bacteria (NOB), which utilize oxygen for the oxidation of nitrite to nitrate (Table 2). As NOB metabolism utilizes nitrite as the metabolic substrate under aerobic conditions, partial nitritation systems represent viable environments for NOB to develop. The most common NOB species isolated from activated sludge systems belong to the genera, *Rubrivivax*, *Rhodobacter* and *Pseudomonas* [32,33]. NOB population development is the major problem related to partial nitritation operational performance. It has been stated that the achievement of the desired ammonium oxidation of a partial nitritation bioreactor depends on the understanding of the AOB and NOB community structure and the effect of operational conditions on AOB and NOB community dynamics [34]. If NOB communities are uncontrolled, nitrate will appear, due to complete nitritation of ammonium when ammonium loading declines from 0.5 kg N m^{-3} day^{-1}, with the consequent loss in the performance of the system. Therefore, controlling the NOB population is needed in order to achieve the desired nitrogen removal [11,22]. The different characteristics of AOB and NOB have been studied in order to develop strategies for NOB control in partial nitritation systems. These are based on temperature, dissolved oxygen, hydraulic retention time (HRT) and solid retention time (SRT), or free ammonia (FA) and free nitrous acid (FNA) concentrations, among others [14,24,35,36]. When 50% ammonium is oxidized under steady-state conditions, partial nitritation reactors have a relative abundance of 64% AOB and less than 5% NOB [37].

EFFECT OF OPERATIONAL CONDITIONS AND QUINOLONE ANTIBIOTIC OVER THE PARTIAL NITRITATION PROCESS

Effect of Influent Characteristics

Partial nitritation reactors have been developed for the treatment of wastewater with low organic matter content and high concentrations of ammonium (Figure 2). The effects that different substances present in wastewater, such as ammonium and organic matter, have on the partial nitritation process have been extensively investigated.

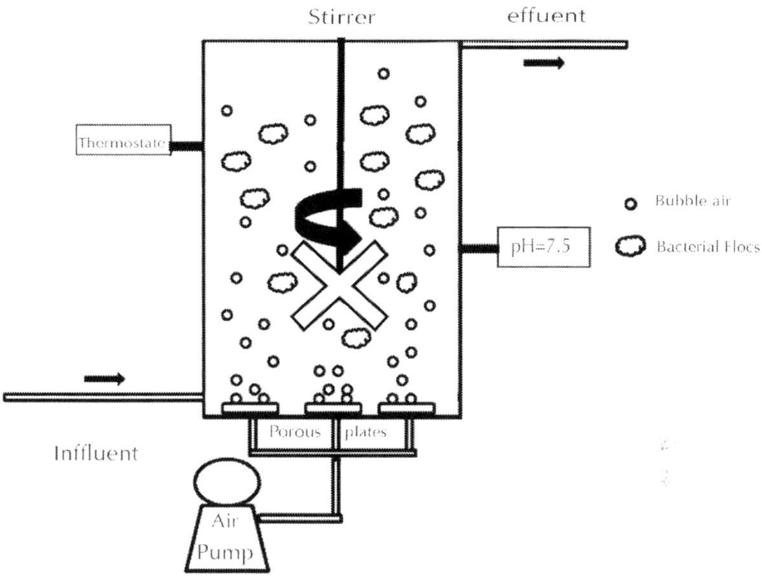

Figure 2: Schematic of a fluidized bed partial nitritation reactor.

The ammonia loading rate (ALR) has been shown to affect the chemical composition of the effluent generated in a partial nitritation reactor. In a continuously-stirred tank reactor (CSTR) configuration, at an ALR from 1.04 $N\text{-}NH_4^+\ m^{-3}day^{-1}$ to 1.81 $N\text{-}NH_4^+\ m^{-3}\ day^{-1}$, the composition of the effluent was stable and dominated by nitrite and ammonia, at a ratio of about 1.2:1, with a small fraction of nitrogen present as nitrate. A similar trend was observed in a biofilm reactor configuration at the ALR range of 2.15 to 4.07 $N\text{-}NH_4^+\ m^{-3}\ day^{-1}$. This effluent composition was suitable for a further anammox treatment for nitrogen elimination. Lower ALRs lead to an excess of nitrate in the effluent for this purpose, and higher ALRs achieve an excess of ammonia in the effluent. In this way, ALR has been proposed as a practical way to control the performance of partial nitritation reactors and has been claimed to be more practical than other control strategies, such as dissolved oxygen (DO) control [38].

Ammonium concentration has been shown to affect the performance of partial nitritation reactors, due to the production of FA and FNA. Both AOB and NOB can be inhibited by their metabolic substrates and/or by-products. It has been found that FA and FNA can inhibit AOB and NOB [39,40,41]. In any case, NOB are much more sensitive than

AOB. The activity of NOB becomes inhibited at FNA concentrations from 0.26 mg HNO_2-N L^{-1} [42], which is lower than the inhibition concentration for AOB, 0.49 mg HNO_2-N L^{-1} [41]. Inhibition by FNA is related to the donation of a proton to the electron transport chain, which impedes the transmembrane pH gradient for the synthesis of ATP [21]. NOB is inhibited by FA at concentrations ranging from 1 to 7 mg NH_3-N L^{-1}, while AOB starts to be inhibited at 150 mg NH_3-N L^{-1} [11]. Other authors have proposed different thresholds for the FA inhibition of NOB (1.75 mg NH_3-N L^{-1}) [42] and AOB (605 mg NH_3-N L^{-1}) [41]. Inhibition of NOB by FA is thought to result from the competition of FA with nitrite oxide reductase (NOR), an NOB enzyme involved in the oxidation of nitrite to nitrate [21]. It has been stated that AOB become inhibited by FA and FNA when nitrogen loading rates become higher than 1.5 kg/m^{-3} day^{-1} [43]. Regardless of this, the adaptation of NOB and AOB communities to FA concentrations has been widely covered by several authors [44,45,46]. Some *Nitrobacter* spp. strains, typical NOB in partial nitritation reactors, have been found to resist up to 40 mg NH_3-N L^{-1} FA [11], while AOB communities have been acclimated to FA concentrations of 122 –224 mg L^{-1} [24]. Therefore, control based on FA concentrations might not be an efficient, practical tool for the assessment of the performance of partial nitritation systems.

Hydroxylamine, an intermediate in the oxidation of ammonium to nitrite, was reported to have the capability of stabilizing a partial nitritation system operating at a high chemical oxygen demand to nitrogen (COD/N) ratio, low temperature and high DO concentration, due to its inhibitory effect on NOB populations [47]. Hydroxylamine has an inhibitory effect on NOB communities at 250 μM and on AOB populations at 2,000 μM; therefore, it selects for AOB to thrive in the system [11].

A certain amount of organic matter can enter a partial nitritation reactor with the influent. Organic matter entering a partial nitritation reactor affects its performance. One of the reasons is that organic matter favors the development of heterotrophs, which have a shorter duplication time than AOB and could therefore outcompete them for oxygen inside the bioreactor [48]. It has been reported that the stability of the partial nitritation process is disturbed by high COD/N ratios, due to the promotion of heterotrophic bacteria inside the system [49]. It has been shown that the C/N ratio does not affect the performance of the partial nitritation process at ammonium volumetric loading rates of 0.5 kg N-NH_4^+ m^{-3} day^{-1}. At higher ammonium volumetric loading rates, higher C/N ratios require higher DO concentrations if the same ammonium oxidation efficiency is desired [24]. In spite of the promotion of heterotrophic growth, the impact of total organic carbon (TOC) concentration on a partial nitritation system has been found to depend on carbon concentration. It has been reported that TOC concentrations are 0.2 g TOC L^{-1}, as acetate stimulates ammonium oxidation in partial nitritation reactors, but also 0.3 g TOC L^{-1}, as acetate decreases 10% ammonium conversion in these systems [6].

Another impact of organic matter on partial nitritation reactors is related to the carbonate-bicarbonate-carbon dioxide equilibrium. With the development of heterotrophs in the partial nitritation reactor, carbon dioxide is generated in the system. Given the pH range in which partial nitritation reactors operate (6.6–8.0), the equilibrium will select for HCO_3 as the dominant form of inorganic carbon. Carbonate will acidify the medium, thus reducing its capacity to balance the release of protons taking place during ammonium oxidation. Therefore, the addition of organic matter will lead to lower conversion rates of ammonium to nitrate [36].

In spite of this fact, it has been proven that the elimination of organic matter can be achieved in partial nitritation reactors at the same time as ammonium oxidation when carbon loading rates do not exceed 2 kg m^{-3} day^{-1} [48]. Accordingly, it has been proposed that, for the purpose of an anammox treatment train, a partial nitritation process following an anammox reactor is recommended when the influent contains a considerable amount of organic matter, which suggests the ability of partial nitritation systems to handle peaks of organic loading while maintaining a stable operational state [50]. In spite of this capacity, it has been found that the recovery of partial nitritation reactors after excessive loading of organic matter is a long process [42].

The addition of certain organic and inorganic compounds could play an important role in partial nitritation processes. It has been found that fulvic acid impacts nitrite accumulation in partial nitritation reactors, impeding ammonium oxidation when its concentration is below 0.002 mg L^{-1} or over 0.07 mg L^{-1}. It has also been reported that NOB are more sensitive than AOB to ortho-cresol, aniline and phenol. ClO_2^- has been proven to inhibit NOB activity at 3 mM; thus, chlorine could be used as a control strategy for NOB population development [21].

Effect of Temperature

Temperature shows a clear relationship with ammonium oxidation in partial nitritation systems. It has been reported that, at a constant ammonium volumetric loading rate and DO concentration, higher temperatures lead to higher ammonium oxidation up to 35 °C. Beyond this point, higher temperatures lead to FA formation, and the activity of AOB becomes inhibited [24]. In similar studies, it was found that at 25 °C, ammonium oxidation reached values of up to 60% of those at 28–39 °C. At 41 °C, bacterial activity stops, and thus, ammonium oxidation does not take place. Furthermore, it has been observed that the ammonium uptake rate in partial nitritation reactors is maximum in the temperature range of 33–37 °C [51].

The bacterial community structure of partial nitritation reactors seems not to be influenced by temperature. Therefore, a difference in microbial activity due to temperature has been proposed as an explanation of differences in the performance of the system at different temperatures [52].

It has been found that NOB have slower growth rates than AOB when temperatures go up from 24 °C, while at temperatures as low as 15 °C, NOB attain domination over AOB in partial nitritation reactors [15,21]. It was reported that temperatures above 30 °C led to a prevalence of AOB over NOB, due to the faster growth rate of the former, so operating at temperatures higher than 30 °C will select for AOB [53].

Furthermore, the effect of temperature on the performance of a partial nitritation reactor has been explained by the formation of FA and FNA, due to shifts in the concentrations of FA and FNA produced in partial nitritation systems. In this way, while the FA concentration increases with temperature, from about 20 mg/L at 25 °C to 120 mg/L at 35 °C, FNA shows the opposite behavior, with 0.5 mg/L at 25 °C and 0.1 mg/L at 35 °C. Thus, at low temperatures, FNA is the main inhibitor of AOB, while at high temperatures, it is the combination of FA and FNA that inhibits ammonium oxidation [54].

In practice, partial nitritation reactors are commonly operated in the range of 30–35 °C to ensure that AOB outcompetes NOB [54]. However, even though partial nitritation reactors have been widely operated at 35 °C, there is not much difference in the practical operation between 25 °C and 35 °C in terms of growth of AOB and NOB; in practical operation, 25 °C is considered enough for the purpose of NOB control [21].

Effect of pH and DO

The pH in a partial nitritation reactor has an impact on its performance. In fact, pH has been reported as a key parameter affecting influent quality in models for laboratory-scale partial nitritation bioreactors [55]. It has been proposed that the influence of pH on ammonium oxidation in partial nitritation reactors is driven by three processes: activation and deactivation of nitrifying enzymes; changes in inorganic carbon concentrations; and changes in FA and FNA concentrations [20,56]. At higher pH, carbonate and bicarbonate are present at higher concentrations, and thus, the buffer capacity of the system increases. On the contrary, as the pH drops below 7.7, the equilibrium tends to carbon dioxide, leading to a loss of buffer capacity [36]. High pH has been related to the formation of FA, which is the primary substrate for AOB, and has also been related to a decrease in FNA concentration, which is the primary substrate of NOB communities [38]. On the other hand, it has been demonstrated that nitrification does not occur below pH 6 [11,20]. Therefore, control of the pH in partial nitritation systems can select for AOB and inhibit NOB, due to the formation of FA and the limitation of FNA [45,57].

It has been reported that the optimal pH for Nitrosomonas spp., typical AOB, ranges between 7.9 and 8.2 and that forNitrobacter spp., typical NOB, it ranges between 7.2 and 7.6, while the optimal pH for partial nitritation reactor operation ranges between 7.0 and 8.0 [11,38]. The higher tolerance of AOB to low pH is thought to result from their ability to develop thick extracellular polymeric substance (EPS) layers [58].

DO concentrations have an impact on the performance of a partial nitritation reactor. It has been found that ammonium oxidation increases with DO concentration, regardless of the C/N ratio, for the same ammonium volumetric loading rate [24].

Half saturation constant values for oxygen of AOB and NOB were reported to be 0.2–0.4 mg L–1 and 1.2–1.5 mg L–1, respectively, which supports the hypothesis of the lower affinity for oxygen of NOB than AOB [11,21]. Following this, oxygen limitation inside partial nitritation reactors is an efficient way to control NOB development. Some studies state that the growth rate of AOB is higher than that of NOB when DO concentrations drop below 1 mg L–1 [22]. Therefore, DO concentrations lower than 1 mg L–1 are used to control NOB in partial nitritation reactors [59,60]. The accumulation of nitrite can be controlled in a short time by setting DO concentration to 0.4–0.8 mg L–1. Furthermore, at DO concentrations of 2 mg L–1 or higher, a substantial accumulation of nitrate occurs in partial nitritation bioreactors [49]. Some authors used aeration of lower than 0.1 (mair3 day–1)/(kg N m–3 day–1) in order to control nitrate accumulation [22]. In addition, low DO concentrations are related to increasing NO and N2O emissions [11].

Partial nitritation processes have been reported as responsible for the emissions of CH_4 and N_2O to the atmosphere [61,62,63,64]. Even though CH_4 emissions have been related to soluble CH_4 stripped from reject wastewater treated in partial nitritation systems, N_2O emissions have been correlated with DO concentration. In this way, higher DO concentrations inside the bioreactor will decrease the $\%N_2O/N_{oxidized}$ ratio. When normal operation parameters were used, emissions of 19.3% ± 7.5%$N_2O/N_{oxidized}$ were found for a partial nitritation sequencing batch reactor (SBR) process [65].

Effect of HRT and SRT

It has been reported that *Nitrosomonas* spp., typical AOB [24], have a maximum growth rate of 0.54 ± 0.09 day^{-1}, while *Nitrobacter* spp., typical NOB [24], have a maximum growth rate of 0.67 ± 0.03 day^{-1} [66]. Other authors have reported similar values for a minimum doubling time of AOB (7–8 h) and NOB (10–13 h) [21]. Given the shorter doubling time of AOB compared with NOB, control of NOB populations can be achieved with the utilization of the HRT/SRT. If HRT is set shorter than the AOB doubling time, then no ammonium oxidation will occur, and if HRT is set longer than the NOB doubling time, then the system will undergo nitrate generation. For this reason, SRT should be set longer than the AOB doubling time, but shorter than the NOB doubling time. In addition to this, partial nitritation bioreactors will avoid sludge retention, given that the recycling of biomass will make NOB persist in the system and, therefore, develop in it. For this reason, partial nitritation reactors are generally operated without sludge retention [8].

Conventionally, HRT and SRT are set up to be the same time in partial nitritation reactors. However, the development of non-coupled HRT and SRT partial nitritation bioreactors has been attempted by biofilter partial nitritation reactors. By setting up a biofilter process, SRT is increased with respect to HRT. It has been confirmed that biotransformation of ammonium to nitrite is three-times higher in biofilter partial nitritation reactors than in conventional partial nitritation reactors [1].

The impact of HRT on partial nitritation reactors has been evaluated by several authors. It was found that higher HRTs, at the same SRTs, led to higher oxidation of ammonium to nitrite [1,25,37]. Differences in ammonium oxidation at different HRTs are due to differences in microbial community structure inside the bioreactor at these HRTs. Differences in bacterial diversity have been reported in biofilter partial nitritation reactors for different HRTs. Within long HRTs of 12 h, the dominant phylotypes are *Nitrosomonas europaea* and *Nitrosomonas eutropha,* with the importance of*Diaphorobacter* sp. At moderately long HRTs of 9 h, the dominant species are *N. europaea, Nitrosospira* spp., and*Paracoccus* spp., with the presence of *Nitrosovibrio* spp., *Rhodobacter* spp. and *Catellibacterium* spp. [1,25]. A decrease in microbial diversity at high HRTs due to competition for ammonium has been observed, which links bacterial community structure to ammonium oxidation in partial nitritation reactors [1]. At high HRTs, the ammonium loading rate decreases in the system. Under these conditions, ammonium is scarce and bacteria with high affinity for ammonium will rapidly thrive inside the bioreactor, outcompeting other bacteria with a lower affinity for this substrate. The selection of bacteria with high ammonium affinity will increase the biotransformation rate of the system. Therefore, control of a biofilter partial nitritation reactor can be achieved by changing the HRT of the system [1].

The SRT also has an influence on bacterial communities in partial nitritation reactors. As a main control of NOB in the system, the SRT should be set lower than the duplication time of NOB, thus ensuring the washout of these bacteria from the reactor [8]. Nevertheless, a short

SRT also leads to a loss of AOB biomass. In this way, partial nitritation reactors have been conventionally operated as suspended growth processes [30,67,68]. On the other hand, it has been confirmed that attached growth partial nitritation processes with attached/granular biomass have advantages over suspended growth partial nitritation processes, such as enhanced AOB biofilm formation [69]. Attached growth processes for partial nitritation systems have been successfully applied at the pilot-plant scale [1,5]. An efficient, stable, 50%–50% ammonium–nitrate ratio in the effluent has been achieved at optimum HRTs of 7 h [5]. In this sense, it can be said that HRTs lower than 7 h could not achieve the oxidation of half of the influent ammonium, with the consequent failure of the operation of the system. An analysis of bacterial community dynamics at shorter HRTs has not been attempted to date in biofilter partial nitritation reactors. It has been proven that attached growth processes offer up to 100-times higher biomass retention and 10-times lower loss of biomass and require a lower HRT than suspended growth processes [70].

Effect of Salinity, Starvation of Biomass and Emerging Contaminants

Salinity affects the performance of partial nitritation systems. It has been concluded that concentrations of 85 mM NaCl increases ammonium conversion by up to 30%. As the NaCl concentration rises, the system loses stimulation and tends to similar values of performance as those for the no-salinity scenario. Nevertheless, at NaCl concentrations of 256 mM and higher, the system loses the capacity for ammonium oxidation. At 342 mM NaCl, ammonium oxidation in the system is 70% less compared with no-salinity operation [4]. The sensitivity of AOB has been identified after short-term exposure to salinity, but the adaptation of *N. europaea* strains to high salinity conditions has also been observed [71,72].

The bacterial dynamics of AOB and NOB communities inside a partial nitritation bioreactor subjected to a long period of starvation were studied [73]. When the feed of a partial nitritation reactor ceases, ammonium availability starts to decrease due to AOB activity, until all of the ammonium has been oxidized. At this point, due to the lack of substrate, AOB communities exhibit a sharp decrease in relative abundance, from 77% to less than 1%. On the other hand, NOB communities gain relative abundance due to the availability of nitrite excreted by AOB. This trend stops when all of the nitrite is consumed within the system. When the starvation period ceases and the bioreactor is subjected to a regular feed again, AOB communities develop rapidly, achieving a steady state similar to that before the starvation period after 15 days.

Emerging contaminants (ECs) have been proposed as particular pollutants that, in fact, have never been studied before [74]. They came to the attention of the scientific community recently and are different from traditional pollutants. Contaminants of emerging concern today include pharmaceutical products, plasticizers, flame retardants, new-generation pesticides, cyanotoxins and more.

Pharmaceuticals have been reported as some of the most important ECs. They are present in urban and industrial wastewater treatment systems all around the world (an example from the analysis of pharmaceuticals from three wastewater treatment plants of Catalonia region, Spain, is given in Table 3) and have been targeted as contaminants of emerging concern [1]. As an example, different antibiotics have been detected at concentrations of up to 450 ng L^{-1} in urban wastewater treatment plants [75]. In this context, it has been reported that the partial nitritation process can be drastically inhibited by quinolone antibiotics, such as ciprofloxacin [76]. Quinolone antibiotics have been found in influent urban wastewater in concentrations ranging up to 426 ng/L [3].

As indicated above, the effect of ciprofloxacin on the performance and the bacterial community dynamics of a partial nitritation reactor has been studied [3]. Ciprofloxacin at 100 ng L^{-1} impacts the performance of the system, which loses 40% capacity of ammonium conversion during the first several days, but stabilizes within a month, showing a more efficient conversion of ammonium than before the antibiotic addition. On the other hand, concentrations up of 350 ng L^{-1} ciprofloxacin drastically reduce ammonium oxidation, by 70%, during the first several days. When the system stabilizes, ammonium conversion recovers, but the performance reached is 30% lower compared with that at the beginning of the experiment. Biomass in the system decreases when ciprofloxacin is added, by about 40% for 100 ng $L^{-1}$and 60% for 350 ng L^{-1}, but at the end of the experiment, there is a net increase in the biomass (5% for the lower concentration and 80% for the higher concentration). In the case of 100 ng L^{-1} ciprofloxacin, bacterial community dynamics start with a decrease of the relative abundance of AOB and end with a slight net increase in the importance of AOB with respect to the starting structure. On the other hand, ciprofloxacin at concentrations of up to 350 ng L^{-1}removes the AOB domination of the system, in favor of *Comamonas* strains. With time, AOB tend to grow in importance; however, the microbial community structure has still shifted completely, and the recovery of the system is impossible (Figure 3).

Table 3: Examples of pharmaceutical emerging contaminants (ECs) detected in influent and effluent wastewater of wastewater treatment plants (adapted from [77]).

Compound	**Frequency of detection (%)**	
	Influent wastewater	**Effluent wastewater**
Carbamazepine	100	100
Diazepam	54	54
Trimethoprim	100	96
Ketoprofen	100	54
Mevastatin	12	8
Clarithromycin	73	85
Salbutamol	69	58
Furosemide	100	96
Sulphamethazine	58	65
Diclofenac	92	100

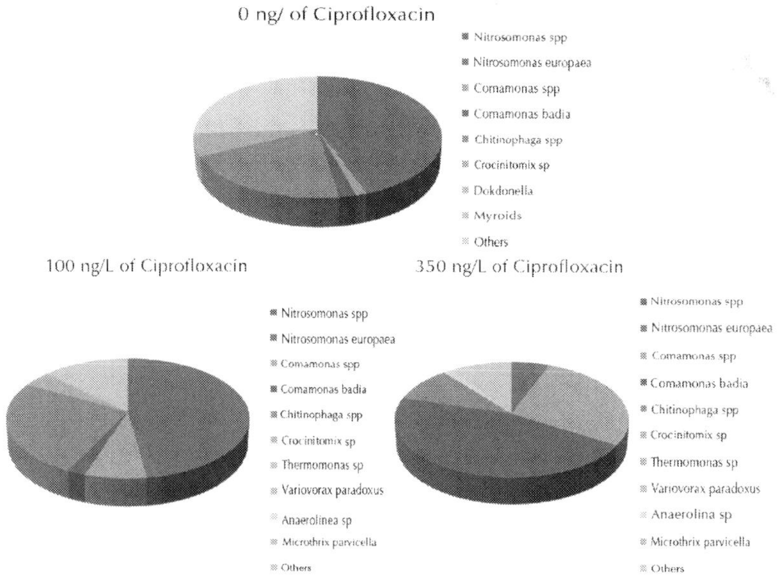

Figure 3: Bacterial community structure of a biofilter partial nitritation bioreactor operating at stable nitrogen removal values subjected to the addition of ciprofloxacin antibiotic to its influent (adapted from [3]). It can be observed that the addition of 350 ng/L of ciprofloxacin dramatically changed the microbial community structure in the system. Furthermore, partial nitritation bioreactors can withstand ciprofloxacin concentrations of up to 100 ng/L with no significant impact on the microbiota.

FUTURE TRENDS FOR THE PARTIAL NITRITATION PROCESS

The application of partial nitritation systems to nitrogen removal from wastewater is becoming more and more popular around the world. Today, full-scale plants that implement the partial nitritation/anammox system for nitrogen elimination from reject wastewater are a reality [3,76]. In addition, the oxidation of ammonium to nitrite is a necessary step for anammox bacteria (Table 2) to carry out autotrophic denitrification [3]. Therefore, the development of systems that rely on the anammox elimination of nitrogen will need to take into account partial nitritation processes.

A promising technology based on anammox microorganisms is the CANON process. In these systems, partial nitritation and anaerobic ammonium oxidation mediated by anammox bacteria take place in the same bioreactor. A coupled partial nitritation-anammox reaction is possible due to the syntrophic metabolism of AOB and anammox bacteria and its spatial organization in granular biomass (Figure 4) [78]. Therefore, the necessity for two different bioreactors for partial nitritation and the elimination of nitrogen via the anammox pathway, as in the partial nitritation/anammox process, is removed. In spite of their engineering advantages, CANON systems are more complicated in practical operation than partial nitritation bioreactors. The competition of NOB with AOB for oxygen and with anammox bacteria for nitrite is the major problem for the performance of these systems [79]. The growth of NOB inside CANON bioreactors is commonly avoided by limiting the DO concentration in the system. The selection of AOB and the inactivation of NOB coupled with the growth of anammox bacteria inside the same bioreactor arises as the immediate future application of partial nitritation.

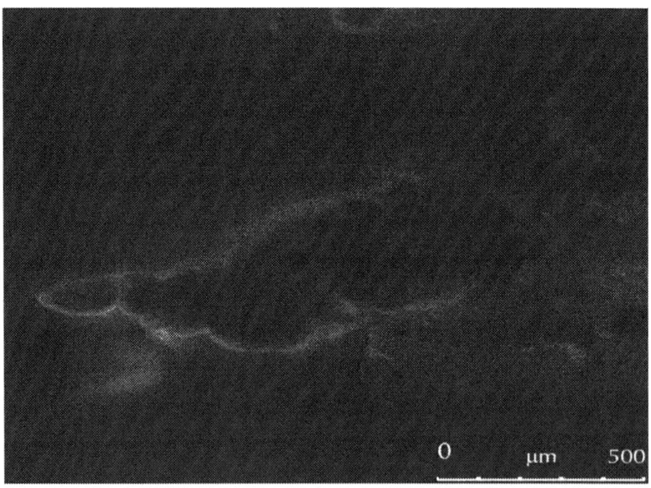

Figure 4: Fluorescence in situ hybridization (FISH) image of a biomass granule from a CANON bioreactor. AOB (green) and anammox bacteria (red) are displaced next to each other; the spatial organization of AOB and anammox bacteria inside the granule is determined by the syntrophic relationships among the phylotypes.

On the other hand, the development of partial nitritation systems has yet to solve the main problem that these processes have: the development of NOB communities inside the bioreactor. Many steps can be taken to further avoid the growth of NOB species in partial nitritation reactors. One of these steps is to turn partial nitritation processes into fully anaerobic processes. This can be done by taking into account the anaerobic nature of autotrophic ammonium oxidation [9].

Strains of *N. eutropha* were reported to perform anaerobic ammonium oxidation with the mediation of nitrogen dioxide as an electron acceptor [80]. Further investigation showed that dinitrogen tetraoxide could also be utilized as the final electron acceptor for anaerobic autotrophic ammonium oxidation in an *N. eutropha* species [81]. Later, a new pathway for anaerobic autotrophic ammonium oxidation was proposed [81]. Here, dinitrogen tetraoxide oxidizes ammonium to hydroxylamine and nitrous oxide with the mediation of the enzyme, AMO. In this way, for the purposes of anaerobic ammonium oxidation in *N. eutropha*, the presence of oxygen just causes nitrous oxide to become oxidized to nitrogen dioxide [82]. Furthermore, the study of the nitrogen cycle in oxygen-depleted marine areas suggests that anaerobic ammonium oxidation is linked to Mn, Fe and S reduction [83,84,85,86].

If AOB can perform anaerobic ammonium oxidation, then partial nitritation reactors could avoid the growth of NOB communities by setting up anaerobic conditions in the system. Further work on technologies relying on this pathway aims to change the future perspectives of autotrophic nitrogen removal processes.

ACKNOWLEDGEMENTS

The authors would like to thank all the Environmental Microbiology and Technical Group of the University of Granada.

AUTHOR CONTRIBUTIONS

Alejandro Rodriguez-Sanchez has been primarily responsible for the compiled information in the review. He has been particularly concerned with the microbiology of AOB, NOB and anammox bacteria

58. Manipura, A.; Duncan, J.R.; Roman, H.J.; Burgess, J.E. Potentia biological processes available for removal of nitrogenou compounds from metal industry wastewater. *Process Saf. Enviror Prot.* 2005, *83*, 472–480.
59. Tokutomi, T. Operation of a nitrite-type airlift reactor at lov DO. *Water Sci. Technol.* 2004, *49*, 81–88.
60. Sinha, B.; Annachhatre, A.P. Partial nitrification—Operationa parameters and microorganisms involved. *Rev. Environ. Sci BioTechnol.* 2006, 6, 285–313.
61. Kampschreur, M.J.; van der Star, W.R.L.; Wielders, H.A.; Mulder, J.W.; Jetten, M.S.M.; van Loosdrecht, M.C.M. Dynamics of nitric oxide and nitrous oxide emission during full-scale reject water treatment. *Water Res.* 2008, *42*, 812–826.
62. De Graaff, M.S.; Zeeman, G.; Temmink, H.; van Loosdrecht, M.C.M.; Buisman, C.J.N. Long term partial nitritation of anaerobically treated black water and the emission of nitrous oxide. *Water Res.* 2010, *44*, 2171–2178.
63. Desloover, J.; de Clippeleir, H.; Boeckx, P.; Du Laing, G.; Colsen, J.; Verstraete, W.; Vlaeminck, S.E. Floc-based sequential partial nitritation and anammox at full scale with contrasting N_2O emissions. *Water Res.* 2011, *45*, 2811–2821.
64. Law, Y.; Ni, B.-J.; Lant, P.; Yuan, Z. N_2O production rate of an enriched ammonia-oxidising bacteria culture exponentially correlates to its ammonia oxidation rate. *Water Res.* 2012, *46*, 3409–3419.
65. Pijuan, M.; Torà, J.; Rodríguez-Caballero, A.; César, E.; Carrera, J.; Pérez, J. Effect of process parameters and operational mode on nitrous oxide emissions from a nitritation reactor treating reject wastewater. *Water Res.*2014, *49*, 23–33.
66. Blackburne, R.; Yuan, Z.; Keller, J. Partial nitrification to nitrite using low dissolved oxygen concentration as the main selection factor. *Biodegradation* 2008, *19*, 303–312.
67. Sliekers, A.O.; Third, K.; Abma, W.; Kuenen, J.; Jetten, M.S. CANON and Anammox in a gas-lift reactor. *FEMS Microbiol. Lett.* 2003, *218*, 339–344.

68. Li, H.; Zhou, S.; Huang, G.; Xu, B. Partial nitritation of landfill leachate with varying influent composition under intermittent aeration conditions. *Process. Saf. Environ. Prot.* 2013, *91*, 285–294.

69. Wik, T. Trickling filters and biofilm reactor modelling. *Rev. Environ. Sci. Biotecnol.* 2003, *2*, 193–212.

70. Choi, J.; Ahn, Y. Comparative performance of air-lift partial nitritation processes with attached growth and suspended growth without biomass retention. *Environ. Technol.* 2014, *35*, 1328–1337.

71. Tal, Y.; Watts, J.E.; Schreier, S.B.; Sowers, K.R.; Schreier, H.J. Characterization of the microbial community and nitrogen transformationprocessesassociatedwithmovingbedbioreactorsin a closed recirculated mariculture system. *Aquaculture* 2003, *215*, 187–202.

72. Moussa, M.S.; Fuentes, O.G.; Lubberding, H.J.; Hooijmans, C.M.; van Loosdrecht, M.C.M.; Gijzen, H.J. Nitrification activities in full-scale treatment plants with varying salt loads. *Environ. Technol.* 2006, *27*, 635–643.

73. Torà, J.A.; Lafuente, J.; Baeza, J.A.; Carrera, J. Long-term starvation and subsequent reactivation of a high-rate partial nitrification activated sludge pilot plant. *Bioresour. Technol.* 2011, *102*, 9870–9875.

74. Sauvé, S.; Desrosiers, M. A review of what is an emerging contaminant. *Chem. Cent. J.* 2014, *8*.

75. Sáez, F.; Pozo, C.; Gómez, M.A.; Rodelas, B.; Gónzalez-López, J. Growth and nitrite and nitrous oxide accumulation of *Paracoccus denitrificans* ATCC 19367 in the presence of selected pesticides. *Environ. Toxicol. Chem.* 2003,*22*, 1993–1997.

76. Van der Star, W.R.L.; Abma, W.R.; Blommers, D.; Mulder, J.-W.; Tokutomi, T.; Strous, M.; van Loosdrecht, M.C.M. Startup of reactors for anoxic ammonium oxidation: Experiences from the first full-scale anammox reactor in Rotterdam. *Water Res.* 2007, *41*, 4149–4163.

77. Jelic, A.; Gros, M.; Ginebreda, A.; Cespedes-Sanchez, R.; Ventura, F.; Petrovic, M.; Barcelo, D. Occurrence, partition and removal of

pharmaceuticals in sewage water and sludge during wastewater treatment. *Water Res.*2011, *45*, 1165–1176.

78. Nielsen, M.; Bollmann, A.; Sliekers, O.; Jetten, M.; Schmid, M.; Strous, M.; Revsbech, N.P. Kinetics, diffusional limitation and microscale distribution of chemistry and organisms in a CANON reactor. *FEMS Microbiol. Ecol.*2005, *51*, 247–256.
79. Liu, T.; Li, D.; Zeng, H.; Li, X.; Zeng, T.; Chang, X.; Zhang, J. Biodiversity and quantification of functional bacteria in completely autotrophic nitrogen-removal over nitrite (CANON) process. *Bioresour. Technol.* 2012, *118*, 399–406.
80. Schmidt, I.; Bock, E. Anaerobic ammonia oxidation with nitrogen dioxide by *Nitrosomonas eutropha*. *Arch. Microbiol.* 1997, *167*, 106–111.
81. Schmidt, I.; Bock, E. Anaerobic ammonia oxidation by cell-free extracts of *Nitrosomonas eutropha*. *Antonie Van Leeuwenhoek* 1998, *73*, 271–278.
82. Schmidt, I.; Bock, E.; Jetten, M.S. Ammonia oxidation by *Nitrosomonas eutropha* with NO(2) as oxidant is not inhibited by acetylene. *Microbiology* 2001, *147*, 2247–2253.
83. Hulth, S.; Aller, R.C.; Gilbert, F. Coupled anoxic nitrification/manganese reduction in marine sediments. *Geochim. Cosmochim. Acta* 1999, *63*, 49–66.
84. Luther, G.W.; Popp, J.I. Kinetics of the abiotic reduction of polymeric manganese dioxide by nitrite: An anaerobic nitrification reaction. *Aquat. Geochem.* 2002, *8*, 15–36.
85. Zehr, J.P.; Ward, B.B. Nitrogen cycling in the ocean: New perspectives on processes and paradigms. *Appl. Environ. Microbiol.* 2002, *68*, 1015–1024.
86. Clement, J.-C.; Shrestha, J.; Ehrenfeld, J.G.; Jaffe, P.R. Ammonium oxidation coupled to dissimilatory reduction of iron under anaerobic conditions in wetland soils. *Soil Biol. Biochem.* 2005, *37*, 2323–2328.

Chapter 4

Sorption of Emerging Organic Wastewater Contaminants to Four Soils

Sarah Roberts, Christopher Higgins and John McCray

Colorado School of Mines, Golden, CO 80401, USA;

ABSTRACT

Conventional onsite wastewater treatment system design relies on a septic tank and soil treatment unit (STU) for treatment of wastewater and integration of the final effluent into the environment. Organic water contaminants (OWCs), chemicals found in pharmaceutical drugs, detergents, surfactants, and other personal care and cleaning products, have been observed in septic tank effluent and the environment. Sorption of OWC mass to soil is a key mechanism in the removal and retardation of many of these chemicals in effluent as it travels through an STU. The primary purpose of this study was to investigate the relationship between the fraction of organic carbon of soil and the equilibrium sorption partitioning coefficient of a selected group of relevant and diverse OWCs. A secondary goal is to evaluate current methods of modeling the sorption of selected OWCs in soil. Five point Freundlich isotherms were constructed from equilibrium sorption batch tests for target OWCs with four different soils. For soils with organic carbon fraction between 0.021 and 0.054, K_d values were calculated between 60 and 185 for 4-nonylphenol, 75 to 260 for triclosan, 115 to 270 for bisphenol-A, 3 to 255 for 17β-estradiol, 40 to 55 for 17α-ethynylestradiol, and 28 to 70 for estrone. An empirically derived, direct relationship between f_{oc} and K_d may be a useful approach to estimating sorption for a soil based on organic carbon content.

INTRODUCTION

Approximately one fifth of the US population is served by onsite wastewater treatment systems (OWTSs), which are a significant and growing means of wastewater treatment in the US [1]. Conventional designs rely on a soil treatment unit (STU) for additional treatment of engineered-unit effluent (such as septic tank effluent) prior to discharge to the environment. Organic wastewater contaminants (OWCs) include a suite of wastewater chemicals such as pharmaceuticals, surfactants, and other personal care products. While the presence of OWCs in OWTS effluent has been well documented [2,3], to date, the factors dictating the efficacy of STUs to remove OWCs have not been comprehensively characterized.

OWCs can have a wide array of ecotoxicological effects and negative impacts on environmental systems. Endocrine disrupting chemicals (EDCs) are a subclass of OWCs that can impair reproductive development and hormone production in some organisms. This effect has been well documented in fish and amphibians, with some populations experiencing skewed gender ratios and/or feminization of males [4]. Other OWCs can affect the renal system, leading to kidney failure [5], while still others are carcinogenic and/or mutagenic [6]. In addition, antimicrobial OWCs may contribute to the evolution of resistant strains of bacteria [7,8,9]. Given these potential adverse impacts and the large and growing fraction of US wastewater treated by

STUs, an accurate understanding of the factors controlling the removal of OWCs in STUs is needed. Moreover, this knowledge will aid in the understanding, design and operation of STUs so as to minimize environmental impacts.

One of the major factors controlling the attenuation of OWCs in STUs is removal via sorption to soil media. Building from traditional organic contaminant partitioning paradigms, most sorption models for OWCs assume that the soil-water partitioning coefficients are related to the organic carbon content (*i.e.*, f_{oc}) of the soil material (Equation (1)) [10,11,12,13,14,15]:

$$K_d = K_{oc} \times f_{oc} \quad (1)$$

where K_{oc} is the organic carbon-normalized partition coefficient, and K_d is the soil-water partitioning coefficient. These K_{oc} values can in turn be calculated from the octanol-water partition coefficient (K_{ow}) using various linear free energy relationships [16,17,18].

For this study, a representative suite of OWCs was selected to examine whether this modeling paradigm is appropriate for sorption of OWCs to soils typical of those used in STUs. This suite includes triclosan (TCS), 4-nonylphenol (NP), bisphenol-A (BPA), estrone (E1), 17β-estradiol (E2), and 17α-ethynylestradiol (EE2). These target OWCs are ubiquitous in wastewater effluent at significant concentrations [2,19,20] and have demonstrated ecotoxicological effects [4,7,8,21,22,23,24,25].

The overarching goal of this study was to investigate the relationship between the f_{oc} of a soil and the K_d of a selected group of relevant and diverse OWCs. In addition, data describing the sorption of these OWCs to a variety of soil types was developed, and current methods of modeling the sorption of the selected OWCs in soil were briefly evaluated.

MATERIALS AND METHODS (OR EXPERIMENTAL)

Chemicals and Soil

Source information and Chemical Abstract Services Registry Number (CASRN) for the six OWCs selected for this study are included in Table 1.

Table 1: Chemical source, Chemical Abstract Services Registry Number (CASRN), Gas Chromatography Mass Spectrometry (GC/MS) retention times, and chemical indicator ions, with qualifying ions used for peak identification in parentheses

Chemical	Source	CASRN	Retention Time (min)	Qualifying Ions (m/z)
Triclosan	KIC Chem, New Paltz, NY, USA	3380-34-5	16.682	289.9 (218.0, 146.0)
4-Nonylphenol	Aldrich, St. Louis, MO, USA	25154-52-3	8.22–12.0	135.1 (220.2, 107.1, 121.1)
Bisphenol A	Aldrich, St. Louis, MO, USA	80-05-7	17.610	213.1 (228.1, 119.0)
Estrone	Aldrich, St. Louis, MO, USA	53-16-7	23.931	270.2 (146.1, 185.1, 172.1)
17b-Estradiol	Aldrich, St. Louis, MO, USA	50-28-2	24.106	272.2 (160.1, 213.1)
17a-Ethynylestradiol	Aldrich, St. Louis, MO, USA	57-63-6	24.993	213.1 (296.2, 160.1, 133.1)
4-normal-nonylphenol	Ehrenstorfer, Augsburg, Germany	104-40-5	12.785	107.05 (220.2, 77.0)
d_6 Bisphenol A	CIL, Tewksbury, MA, USA	86588-58-1	17.551	216.1 (234.1, 121.1)
d_4 17β-estradiol	CIL, Tewksbury, MA, USA	66789-03-5	24.084	276.2 (161.1, 147.1, 214.1)

Four soils were obtained from locations in Colorado, Nevada, and North Dakota, and bulk soil parameters were measured by a commercial laboratory for each soil type and included: pH, organic carbon content (f_{oc}), particle size distribution, and cation exchange capacity (Table 2).

Table 2: Soil characteristics for the four soils considered in sorption batch tests. CEC is the cation exchange capacity

Characteristic	Sand	Loamy Sand	Sandy Loam	Loam
Organic Carbon (%)	0.1	2.1	3.1	5.4
Soil pH	10.4	5.8	6.7	6.9
CEC (cmol(+)/kg)	6.0	12.0	11.3	26.4
Ca (% of cations)	86.6	54.2	73.3	66.2
Mg (% of cations)	10.9	12.5	19.6	16.7
K (% of cations)	1.0	10.2	6.6	5.4
Na (% of cations)	1.4	1	0.5	0.4
% Sand	98	86	65	30
% Silt	0.8	6	15	44
% Clay	1.2	8	20	26

Experimental Setup

Sorption isotherms were developed for the six OWCs in the four soils using the aqueous-loss batch approach. Simulated fresh water composed of 0.01 M calcium chloride ($CaCl_2$) was used with a 1:5 soil-water mass ratio in 40 mL glass centrifuge tubes for all replicates and concentrations. This ratio was successfully used in previous experiments involving similar soil [2] and in sorption experiments using the same analytes and similar soil [26]. Prior to initiation of the isotherm experiments, the four soils (sand, loamy sand, sandy loam, and loam) were autoclaved twice for 45 min at 121 °C and 103 kPa to inactive any microbes present and minimize losses due to biodegradation. Preliminary studies comparing sterilization methods found no significant differences in sorption for autoclaved soil when compared to 5% sodium azide biocide solutions (data not shown). While extraction of solids to close the mass balance was not performed, given the short equilibration time and the use of autoclaved soil, minimal biotransformation of the target analytes was expected. Background concentrations of analytes in soils were below analytical detection limits. While pH was not directly measured in the isotherms, the pH of these batch systems was assumed to be similar to the soil pH measured (at an identical 1:5 soil-water ratio; Table 2) by the commercial laboratory.

Batch experiments were conducted in triplicate at five initial concentrations ranging from 60 to 2000 μg/L. This range was chosen based on concentrations of analytes observed in onsite wastewater influent and effluent [2] and analytical method sensitivity. After addition of the approximately 6 g of soil and 30 mL of synthetic freshwater to each tube, samples were mixed with a vortex and then agitated for 36 h on a shaker table (130 rpm). Sorption kinetics were not evaluated: several studies using a similar experimental protocol have reported that equilibrium between the aqueous and sorbed phases was achieved within 22 to 36 h [11,27,28]. After 36 h, glass centrifuge tubes were removed and centrifuged for 20 min at 7000 rpm (3600 rcf) to separate suspended soil particles from solution. The supernatant was decanted and three chemical surrogates (d6-bisphenol-A, 4-normal-nonylphenol, and d4-17β-estradiol) were added to samples at known concentrations to evaluate aqueous analyte concentrations.

Aqueous Phase OWC Analysis

Target OWCs were extracted from aqueous samples using a method adapted from Barber *et al.* (2007) [21]. Prepackaged solid phase extraction (SPE) cartridges (Waters Sep-Pak® Plus tC18) were preconditioned with dichloromethane (DCM) using an initial flush of approximately 100 pore volumes. Samples were then loaded on the SPE cartridges under vacuum suction maintaining low flow (approximately 1 mL per minute). Finally, the target OWCs were eluted with 5 mL of DCM under similar low flow conditions. Anhydrous sodium sulfate was used to ensure all water was removed from the vials. The remaining volume was evaporated to 0.5 mL under ultra-high purity nitrogen gas and transferred to a 1 mL glass GC vial Sample analysis was conducted in triplicate and two samples spiked with analytes underwent extraction processes for every run to evaluate extraction efficiencies. Recoveries were routinely greater than 99%. Quality assurance and quality control procedures also included triplicate analysis of extractions and use of known concentrations to evaluate the mass recovery capability of the SPE method.

Sample analysis was conducted via electron impact gas chromatography (GC) coupled with selected ion monitoring mass spectrometry (MS). A Hewlett Packard (HP) 6890 GC fitted with an

HP Ultra II (5% phenylmethyl silicone) column (25 m by 0.2 mm, 33 μm) was employed with ultra high purity helium as the carrier gas. The injection port was maintained at 300 °C, while the initial oven temperature was 140 °C and increased with a ramp rate of 6 °C per minute up to the final temperature of 300 °C, at which point it was held constant for 15 min. The MS (HP 5793 Mass Selective Detector) was operated with an ionization energy of 70 eV, a source temperature of 250 °C, and an interface temperature of 300 °C. Concentrations of analytes were calculated based on selective ion monitoring abundance data using diagnostic ions for each chemical relative to abundance data for the respective surrogates. These ions are identified by mass and ion retention time determined from standard analysis and a peak signal to noise ratio of 3:1 (Table 1).

Sorption Coefficients

The Freundlich isotherm model was used to describe the sorption of OWCs from solution [29]. The form of the equation employed was:

$$C_s = K_f \times C_w^{\,n} \qquad (2)$$

where: C_s is the concentration of the chemical adsorbed to soil particles (μg/kg), K_f is the Freundlich sorption coefficient (L/kg), C_w is the concentration of chemical in the aqueous phase (μg/L), and n is a power function related to the nonlinearity of the isotherm (n = 1 indicates a linear isotherm).When n is unity, the linear isotherm can be described by:

$$C_s = K_d \times C_w \qquad (3)$$

Where K_d is the sorption coefficient (L/kg).

If the relationship between the magnitude of sorbed concentration and the magnitude of aqueous concentration is linear (*i.e.*, n = 1), then the K_d value is independent of the concentration. However, if the isotherm is nonlinear and a comparison between different sorbents is desired, interpolation of K_d values from the K_f and n values for a specific concentration is appropriate. In addition, if the partitioning of the OWC aqueous phase to the soil is due to the hydrophobic sorption of the contaminant to the f_{oc} of the soil, an organic carbon partitioning coefficient (K_{oc}) can be calculated with Equation (1). Linear regressions were evaluated using the sum of least squares-method to find the straight line of best fit with Microsoft Excel.

RESULTS AND DISCUSSION

Sorption of OWCs on Different Soils

The isotherms for the batch sorption isotherms for these experiments were generally well-described by the Freundlich isotherm model (Equation (2), Figure 1), and linear in nature (*i.e.*, the 95% confidence intervals for the Freundlich exponent (*n*) values included unity). In general, K_d values for all targeted OWCs (Table 4) were much lower in the sand than for all other soils and were generally higher for soils with greater organic carbon content. These values agree with those reported by Karnjanapiboonwong *et al.* (2010) [30].

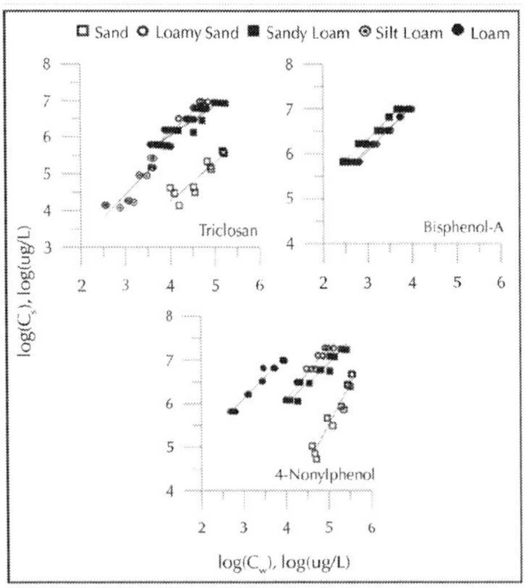

Figure 1: Sorption isotherms for TCS, BPA, and NP for soil isotherm data with an R^2 value greater than 0.80.

A notable exception to the linear sorption behavior was NP sorption onto the sandy soil, which exhibited highly nonlinear sorption (Figure 2). Due to the high variability of the results for some analytes in some soils, only data sets whose Freundlich model fits had R^2 value greater than 0.8 were included in this analysis. Efforts to fit these data with other models (*i.e.*, Langmuir) did not result in better fits. For the sandy soil, the n value for NP was 1.87 ± 0.14 (fitted n ± 95% CI), suggesting that the ratio of sorbed NP mass to aqueous mass increased for higher NP concentrations for this soil. Figure 2 displays equilibrium aqueous concentrations against sorbed concentrations for NP in the sandy soil and also in the sandy loam for comparison. The nonlinear trend is apparent in this plot.

This sorptive behavior has been observed previously for other surfactants, and on a mechanistic level, can be attributed to the hydrophobic self-attraction between the aliphatic carbon chain tails of the NP molecule [31,32]. As is evident in Table 2, the f_{oc} in the sandy soil was quite low (0.001). Because sorption of NP increases with increasing aqueous NP levels, it is possible that the chemical is attracted to its own sorbed phase and that this was the dominant sorption mechanism because of the low f_{oc} Sorption of NP to the sandy soil was quite low, and the K_f value extrapolated from the Freundlich fit is between 0 and 0.002 L/kg considering the 95% confidence intervals. This is at least three orders of magnitude lower than any other K_f value calculated for any chemical in any soil in the present study. It is possible that in soils where the f_{oc} was greater, the sorption to the organic carbon masked this self-sorbing effect, resulting in linear isotherms where the equilibrium between NP in the dissolved and solid phases is primarily dependent on the f_{oc} of the soil.

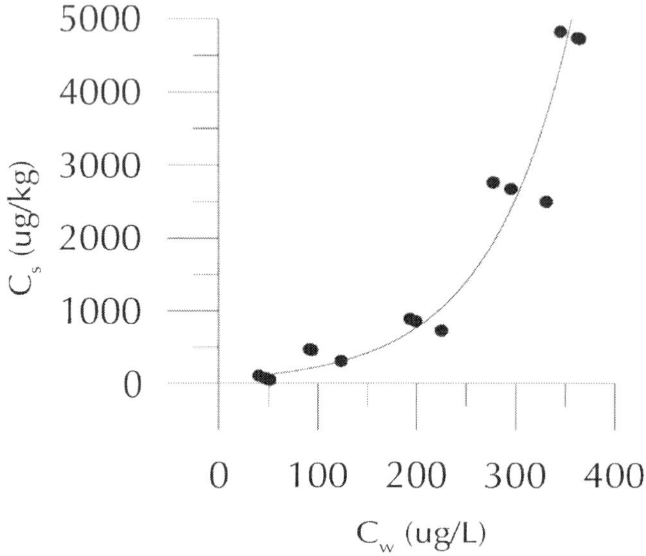

Figure 2: Plots of equilibrium aqueous concentrations and sorbed concentration for 4-nonylphenol in sand.

Relationship between OWC Sorption and Fraction of Organic Carbon

K_d values appear to generally increase with organic carbon content, and though a clear linear relationship was not evident (Figure 3), this may have been due to the limited number of soil types evaluated in the present study. In addition, the lack of a clear linear relationship may also have been due to the uncontrolled solution effects and/or the potential that sorption mechanisms other than partitioning to organic carbon may have also been important for some soils. Examples of other potential sorption mechanisms include ion exchange, surface adsorption to mineral constituents, and complex formation with metal ions. Solution characteristics, such as dissolved organic carbon content that were not accounted for in this study may influence the degree to which OWCs may sorb. The dissolved organic carbon content of the solution may decrease the sorptive capacity by decreasing the attraction of the hydrophobic molecules to the soil organic carbon.

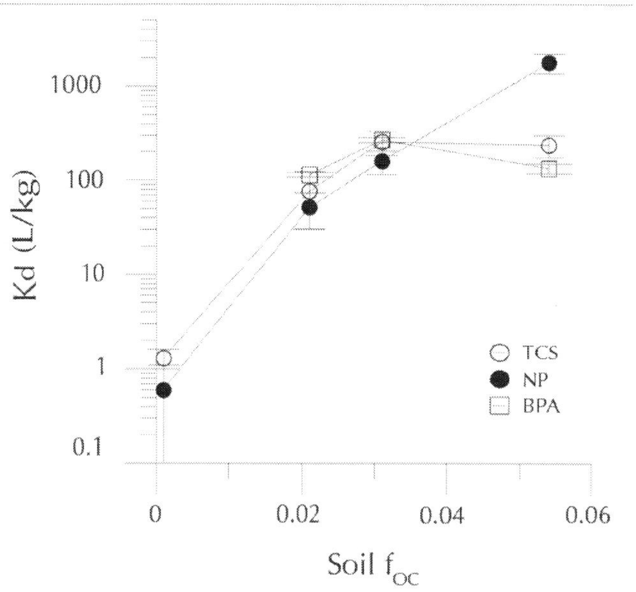

Figure 3: Plots of Kd values and corresponding foc for soil in which Kd values were measured for analytes.

Influence of pH on Sorption of Anionic forms of Aqueous OWCs

Hydrophobic sorption of organic chemicals to organic carbon is commonly thought to depend on the form of the chemical in solution [18,33,34], which, for ionogenic organic chemicals, is dependent on the pH of the system. For many OWCs, the pK_a's are relatively high and outside the normal pH range for wastewater systems, suggesting that pH effects would be minimal. However, for acidic OWCs, dissociation of the neutral form to form the organic anion may occur within the normal pH range of wastewater. Electrostatic repulsion between the organic anions and the typically negatively charged soil particle surfaces may result in substantially weaker sorption. A simple model for describing the effect of pH on the sorption of acidic organic chemicals to natural organic matter in soil is described by Schwarzenbach *et al.* (2003) with Equation (4) [29].

$$D_{oc} = \alpha_a \times K_{OC}^{HA} \qquad (4)$$

Where D_{oc} is the distribution ratio accounting for sorption only of the protonated (neutral) species to natural organic matter, α_a is the fraction of the chemical in the protonated (neutral) form and K_{OC}^{HA} is the sorption coefficient for neutral species. This relationship assumes that the deprotonated (negatively charged) OWC species do not significantly contribute to sorption of the chemical [35,36].

As can be observed in Table 3, the pK_a values of most of the target OWCs considered in this study are relatively high. While no pKa values are reported for estrone or 17α-ethynylestradiol, the assumption can be made that they would be similar to that of 17β-estradiol because of chemical similarities between the three chemicals. The reported pK_a of TCS (8.14) is lower than the measured pH for one soil. If the pH values measured by the commercial laboratory are truly indicative of the pH of the batch systems and the anionic forms of the OWCs do not sorb to the organic carbon, then the resultant K_{oc} values would be substantially lowered for those soils whose pH values fall above the pK_a's for the target OWCs. However, TCS exhibited a relatively high $\log K_{oc}$ (3.12 L/kg) in the sand. While not conclusive, this sorption of presumably anionic TCS agrees with the findings of Tulp *et al.* (2009), Shareef *et al.* (2006) and Butler *et al.* (2012), all of which concluded that anionic forms of organic chemicals are capable of stronger sorption than previously expected [36,37,38].

Comparison of Experimentally Determined and Literature Reported K_{oc} Values

It is illustrative to compare the range of K_{oc} values calculated from the present data to those reported in the literature for these OWCs (Table 3). Literature reported K_{oc} values are grouped into those calculated using structure-activity models, those derived from single-point equilibrium sorption experiments, and those derived from isotherm equilibrium sorption experiments (which are most analogous to the present study). As is clear from Table 4, the range of K_{oc} values reported in the literature spans more than an order of magnitude for most of the chemicals. While some variability might be expected given the range of approaches from which these data are derived, these data suggest that K_{oc} is not a constant value for these chemicals and may be dependent on the system from which K_{oc} values are calculated. Indeed, the range of values reported for these OWCs was one of the motivations for the present study. Table 4 also highlights the fact that relatively few studies have actually measured K_{oc} values for the target OWCs, indicating a gap in current knowledge.

As illustrated in Table 4, the range of K_{oc} values calculated from data in the present study generally agree with those reported previously. The generally large range of reported K_{oc} values suggests that the approach itself may have limited applicability: K_{oc} values tend to range over an order of magnitude for most of the target OWCs. Results from Gerstl *et al.* (1990) indicate that $\log K_{oc}$ values varied by 0.3 in a study of 32 soils and 36 sediments for two OWCs, tetrachloromethane and 1,2-dichlorobenzene [55]. This variability was attributed to the source soil/sediment organic material, which may explain some of the variability in literature reported K_{oc} values. In some cases represented in Table 4, when the f_{oc} of the soil was very low, the reported K_{oc} value may have been distorted due to inaccurate measurements of f_{oc}. For example, the value measured from isotherm batch tests reported for NP ($\log K_{oc}$ = 4.6) was based on sorption to aquifer material with an f_{oc} of 0.005 [46]. Conversely, the $\log K_{oc}$ values reported by Sekela *et al.* (1999), ranging from 4.7 to 5.6, were derived from data for sorption to sewage sludge, which has a relatively high f_{oc} (approaching 0.5) and facilitates a greater degree of hydrophobic sorption than a soil system may potentially achieve [41]. Although K_{oc} values should not depend on the f_{oc}, this comparison shows that the value is clearly not constant for these chemicals. Nevertheless, various fate and transport assessments have utilized a K_{oc} value to reference estimate a K_d value for OWC fate and transport modeling in soil systems [56,57].

Table 3: K_{oc} values from isotherm experiments with R^2 greater than 0.8 presented in this study compared with K_{oc} values reported in the literature. Chemical parameters, pKa and log(K_{ow}) are also listed for analytes. ND designates a value for which no data was reported.

Chemical	Measured Log(K_{oc}) Values		Literature Reported Log(K_{oc}) Values				Chemical Parameters	
	Ave ± SD (N)	Range	Range	Predicted	Measured–Single Point	Measured–Isotherm	pKa	Log (K_{ow})
4 Nonylphenol	3.61 ± 0.73 (4)	2.74–4.52	3.9–5.4	ND	4.7–5.6 [d], 5.3 [e], 4.5–5.2 [f], 4.7 [e]	3.9 [j], 4.6 [k]	10.28 [o]	4.48 [t]
Triclosan	3.56 ± 0.34 (5)	3.12–3.92	3.2–4.6	3.8–4.0 [a], 3.2 [b]	4.6 [g]	ND	7.9 [p]	4.76 [p]
Bisphenol–A	3.69 ± 0.27 (3)	3.40–3.94	2.8–3.9	3.2 [c], 2.8 [b]	ND	3.5 [i], 3.8 [l], 2.9 [m], 3.9 [n]	9.6–11.3 [q]	3.40 [q]
Estrone	3.50 ± 0.16 (3)	3.34–3.65	3.2–3.6	3.6 [b]	3.2 [h], 3.3 [i]	3.3 [l]	10.4 [r]	2.95 [u]
17β-Estradiol	3.23 ± 0.16 (3)	3.11–3.42	3.3–3.6	3.6 [b]	3.3 [i]	3.3 [l]	10.2 [s]	3.86 [u]
17α-Ethynylestradiol	3.25 ± 0.16 (2)	3.13–3.36	2.9–3.3	ND	3.2 [h], 3.3 [i]	2.9 [l]	10.2 [s]	3.67 [u]

Notes: a: [39]; b: [26]; c: [40]; d: [41]; e: [42]; f: [43]; g: [44]; h: [27]; i: [45]; j: [11]; k: [46]; l: [13]; m: [28]; n: [47]; o: [48]; p: [49]; q: [50]; r: [51]; s: [52]; t: [53]; u: [54]; ND: No Data.

The calculation of this partitioning coefficient incorporates a number of assumptions. First, organic carbon is considered as a relatively homogenous phase, essentially meaning that sorption of a chemical does not depend on the type of organic carbon (*i.e.*, root material, leaf litter, decomposing organisms). While studies have shown that the type and geologic age may play a minor role, the soils chosen for the study are analogous to soils used by STUs [58]. Second, the role of dissolved organic carbon in the wastewater stream is not considered in calculation of K_d and K_{oc}. Dissolved organic carbon can serve to reduce the sorption capacity of an OWC to soil and can facilitate solubilization of the compound and aid in transport through the soil. Additionally, because the K_{oc} is calculated using K_d and f_{oc} values, the assumption is made that hydrophobic partitioning is the only sorptive mechanism and this partitioning is defined by a linear relationship (Equation (3)). Nevertheless, because f_{oc} of a soil is an important factor in the degree of sorption of these chemicals, the K_{oc} of a chemical may be useful in predicting the sorptive capacity of an OWC for a given system. Identifying a useful K_{oc} value from literature studies may incorporate assumptions that lead to a lower degree of confidence in fate and transport modeling results. An empirically determined range of K_{oc} values may be a superior approach to estimating a K_d model input value, but is not always possible for every soil due to resource constraints.

CONCLUSIONS

Sorption to soil is a key process in the fate and transport dynamics of OWCs as they infiltrate through soil. The K_d value is used to characterize sorption in fate and transport models that consider the advection-dispersion equation incorporating solute transport and, generally, model output is highly sensitive to the magnitude of this parameter. Clearly, effective characterization of a K_d value and the uncertainty surrounding it for a specific system is essential for fate and transport modeling. To help evaluate this uncertainty, a relationship may be defined to relate a range of K_d values to soil organic carbon content. This range (defined by 95% confidence intervals for a linear regression) incorporates the variability associated with additional soil and solution characteristics by involving a number of various experimentally determined K_d values and not accounting for the variability or effects of soil characteristics other than the f_{oc}. Using this approach, a K_{oc} relationship is defined by considering a number of literature-reported sorption studies over an applicable range of organic carbon contents. Figure 4 illustrates an approach for using this empirical relationship. In this plot, K_d values from this study and literature for E1, E2, and EE2 are combined under the assumption that these chemicals, being similar in structure, demonstrate similar sorption characteristics [13,26,27,45]. A regression was fit between the K_d values and corresponding f_{oc} values and 95 and 99 percent confidence intervals are included. These confidence intervals may define a range of effective K_d values for an f_{oc} based on empirical evidence. It is important to note that the relationship displayed in Figure 4 suggests an approach to identifying a K_d range, but it is unlikely that current K_d information is sufficient to use this approach as a tool for estimating K_d without additional investigation for most OWCs. Rather than requiring extensive numbers of batch studies to characterize K_d values at a variety of soil conditions, a confined range defined by this type of empirical relationship may give a realistic indication as to the effect of sorption on retardation of OWC infiltration.

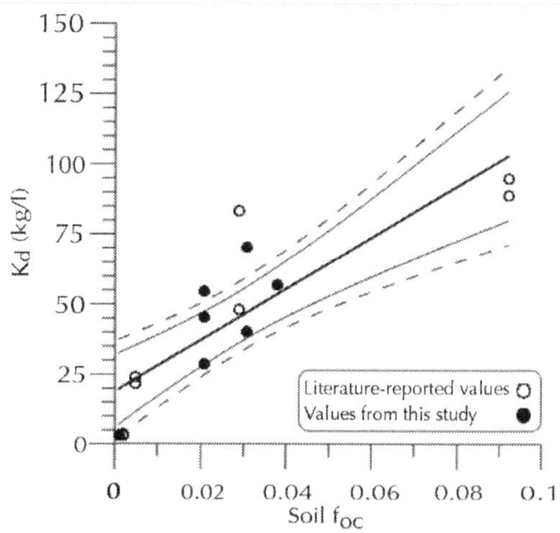

Figure 4: Plots of the Kd values for associated fraction of organic carbon of soil for isotherms with less than 20% variability in this study and literature re-

ported values [13,26,27,45] for estrogenic hormones (estrone, 17**β**-estradiol, and 17**α**-ethynylestradiol). A linear regression fit is included along with the 95% (solid lines) and 99% (dashed lines) confidence intervals.

ACKNOWLEDGMENTS

The authors thank Kathryn S. Lowe (CSM) for an initial review of the manuscript. This work was supported by a research grant from the Water Environment Research Foundation (WERF) (Project DEC 1R06).

AUTHOR CONTRIBUTIONS

Sarah Roberts designed and conducted the experimental work and literature review presented in this paper under the direction, review and guidance of Christopher Higgins and John McCray.

CONFLICTS OF INTEREST

The authors declare no conflict of interest.

REFERENCES

1. U.S. Environmental Protection Agency. *Onsite Wastewater Treatment Systems Manual*; U.S. Environmental Protection Agency: Cincinnati, OH, US, 2002.
2. Conn, K.E.; Barber, L.B.; Brown, G.K.; Siegrist, R.L. Occurrence and fate of organic contaminants during onsite wastewater treatment. *Environ. Sci. Technol.* 2006, *40*, 7358–7366.
3. Matamoros, V.; Arias, C.; Brix, H.; Bayona, J.M. Preliminary screening of small-scale domestic wastewater treatment systems for removal of pharmaceutical and personal care products. *Water Res.* 2009, *43*, 55–62.
4. Brown, A.R.; Riddle, A.M.; Cunningham, N.L.; Kedwards, T.J.; Shillabeer, N.; Hutchinson, T.H. Predicting the effects of endocrine disrupting chemicals on fish populations. *Hum. Ecol. Risk Assess.* 2003, *9*, 761–788.

5. Oaks, J.L.; Gilbert, M.; Virani, M.Z.; Watson, R.T.; Meteyer, C.U.; Rideout, B.A.; Shivaprasad, H.L.; Ahmed, S.; Chaudhry, M.I.; Arshad, M. Diclofenac residues as the cause of vulture population decline in Pakistan. *Nature* 2002, *40*, 65–81.
6. Robertson, W.D. Chemical fate and transport in a domestic septic system: Site description and attenuation of dichlorobenzene. *Environ. Toxicol. Chem.* 1994, *13*, 183–191.
7. Glaser, A. The Ubiquitous Triclosan: A Common Antibacterial Agent Exposed. Beyond Pesticides Fact Sheet. Available online: www.beyondpesticides.org/pesticides/factsheets/Triclosan%20cited.pdf (accessed on 24 March 2004).
8. Singh, L. *An Experiment to Identify Levels of Triclosan That are Harmful to Bacteria in Puget Sound*; University of Washington: Seattle, WA, USA, 2007.
9. Suarez, S.; Dodd, M.; Omil, F.; von Gunten, U. Kinetics of triclosan oxidation by aqueous ozone and consequent loss of antibacterial activity: Relevance to municipal wastewater ozonation. *Water Res.* 2007, *41*, 2481–2490.
10. Das, B.S.; Lee, L.S.; Rao, P.C.; Hultgren, R.P. Sorption and degradation of steroid hormones in soils during transport: Column studies and model evaluation. *Environ. Sci. Technol.* 2004, *38*, 1460–1470.
11. Duering, R. Sorption behavior of nonylphenol in terrestrial soils. *Environ. Sci. Technol.* 2002, *36*, 4052–4057.
12. Duran-Alverez, J.; Prado, B.; Ferroud, A.; Narcedalia, J.; Jimenez-Cisneros, B. Sorption, desorption and displacement of ibuprofen, estrone, and 17β-estradiol in wastewater irrigated and rain-fed agricultural soils. *Sci. Total Environ.* 2014, *473–474*, 189–198.
13. Lee, L.S.; Strock, T.J.; Sarmah, A.K.; Rao, P.C. Sorption and dissipation of testosterone, estrogens, and their primary transformation products in soils and sediment. *Environ. Sci. Technol.* 2003, *37*, 4098–4105.
14. Scheytt, T.; Mersmann, P.; Lindstädt, R.; Heberer, T. Determination of sorption coefficients of pharmaceutically active substances carbamazepine, diclofenac, and ibuprofen, in sandy sediments. *Chemosphere* 2005, *60*, 245–253.

15. Yu, Z.; Xiao, B.; Huang, W.; Peng, P. Sorption of steroid estrogens to soils and sediments. *Environ. Toxicol. Chem.* 2004, *23*, 531–539.

16. Schwarzenbach, R.P.; Westall, J. Transport of nonpolar organic compounds from surface water to groundwater. Laboratory sorption studies. *Environ. Sci. Technol.* 1981, *15*, 1360–1367.

17. Wilson, J.T.; Enfield, C.G.; Dunlap, W.J.; Cosby, R.L.; Foster, D.A.; Baskin, L.B. Transport and fate of selected organic pollutants in a sandy soil. *J. Environ. Qual.* 1981, *10*, 501–506.

18. Zwiener, C.; Frimmel, F.H. Short-term tests with a pilot sewage plant and biofilm reactors for the biological degradation of the pharmaceutical compounds clofibric acid, ibuprofen, and diclofenac. *Sci. Total Environ.* 2003, *309*, 201–211.

19. Lietz, A.C.; Meyer, M.T. Evaluation of emerging contaminants of concern at the south district wastewater treatment plant based on seasonal sampling events, Miami-Dade County, Florida, 2004. In *Scientific Investigations Report 2006–5240*; U.S. Geological Survey: Reston, VA, USA, 2006.

20. Swartz, C.H.; Reddy, S.; Benotti, M.J.; Yin, H.; Barber, L.B.; Brownawell, B.J.; Rudel, R.A. Steroid estrogens, nonylphenol, ethoxylate metabolites, and other wastewater contaminants in groundwater affected by a residential septic system on Cape Cod, MA. *Environ. Sci. Technol.* 2006, *40*, 4892–4900.

21. Barber, L.B.; Thurman, E.M.; Schroeder, M.P.; LeBlanc, D.R. Long-term fate of organic microbpolutants in sewage-contaminated ground water. *Environ. Sci. Technol.* 1998, *22*, 205–211.

22. Howdeshell, K.L.; Hotchkiss, A.K.; Thayer, K.A.; Vandenbergh, J.G.; Vom Saal, F.S. Environmental toxins: Exposure to bisphenol A advances puberty. *Nature* 1999, *401*, 763–764.

23. Lech, J.J.; Lewis, S.K.; Ren, L. *In vivo* estrogenic activity of nonylphenol in rainbow trout. *Fundam. Appl. Toxicol.* 1996, *30*, 229–232.

24. Orvos, D.R.; Versteeg, D.J.; Inauen, J.; Capdevielle, M.; Rothenstein, A.; Cunningham, V. Aquatic toxicity of triclosan. *Environ. Toxicol. Chem.* 2002, *21*, 1338–1349.

25. Roberts, P.; Roberts, J.P.; Jones, D.L. Behavior of the endocrine disrupting chemical nonylphenol in soil: Assessing the risk

associated with spreading contaminated waste to land. *Soil Biol. Biochem.* 2006, *38*, 1812–1822.

26. Ying, G.G.; Kookana, R.S. Triclosan in wastewaters and biosolids from Australian wastewater treatment plants. *Environ. Int.* 2007, *33*, 199–205.
27. Casey, F.M.; Simunek, J.; Lee, J.; Larsen, G.L.; Hakk, H. Sorption, mobility, and transformation of estrogenic hormones in natural soil. *J. Environ. Qual.* 2005, *34*, 1372–1379.
28. Fent, G.; Hein, W.J.; Moendel, M.J.; Kubiak, R. Fate of 14C-bisphenol A in soils. *Chemosphere* 2003, *51*, 735–746.
29. Schwarzenbach, R.P.; Gschwend, P.M.; Imboden, D.M. *Environmental Organic Chemistry*; Wiley-Interscience: New York, NY, USA, 2003.
30. Karnjanapiboonwong, A.; Morse, A.N.; Maul, J.D.; Anderson, T.A. Sorption of estrogens, triclosan, and caffeine in a sandy loam and a silt loam soil. *J. Soils Sediments* 2010, *7*, 1300–1307.
31. Goloub, T.P.; Koopal, L.K.; Bijsterbosch, B.H.; Sidorova, M.P. Adsorption of cationic surfactants on silica. Surface charge effects. *Langmuir* 1996, *12*, 3188–3194.
32. Xiaoping, L.; Laixiang, Z.; Linlin, Y.; Jiale, L.; Liang, X; Mulindankaka, E. Sorption behavior of nonylphenol (NP) on sewage-irrigated soil: Kenetic and thermodynamic studies. *Sci. Total Environ.* 2012, *473–474*, 520–536.
33. Scheytt, T.J.; Mersmann, P.; Heberer, T. Mobility of pharmaceuticals carbamazepine, diclofenac, ibuprofen, and propyphenazone in miscible-displacement experiments. *J. Contam. Hydrol.* 2006, *83*, 53–69.
34. Tsai, W.; Chi-Wei, S.; Ting-Yi, S. Adsorption of bisphenol-A from aqueous solution onto minerals and carbon adsorbents. *J. Hazard. Mate.* 2006, *134*, 169–175.
35. Escher, B.I.; Scharzenbach, R.P. Partitioning of substituted phenols in liposome−water, biomembrane−water, and octanol−water systems. *Environ. Sci. Technol.* 1995, *30*, 260–270.
36. Tulp, H.C.; Fenner, K.; Schwarzenbach, R.P.; Goss, K.U. pH-dependent sorption of acidic organic chemicals to soil organic matter. *Environ. Sci. Technol.* 2009, *43*, 419–445.
37. Shareef, A.; Angove, M.J.; Wells, J.D.; Johnson, B.B. Sorption

of bisphenol A, 17 -ethynylestradiol and estrone to mineral surfaces. *J. Colloid Interface Sci.* 2006, *297*, 62–69.

38. Butler, E.; Whelan, M.J.; Sakrabani, R.; van Egmond, R. Fate of triclosan in field soils receiving sewage sludge. *Environ. Pollut.* 2012, *167*, 101–109.
39. Lindstroem, A.; Buerge, I.J.; Poiger, T.; Bergqvist, P.A.; Mueller, M.D.; Buser, H.R. Occurrence and environmental behavior of the bactericide triclosan and its methyl derivative in surface waters and in wastewater. *Environ. Sci. Technol.* 2002, *36*, 2322–2329.
40. Staples, C.A.; Dome, P.B.; Klecka, G.M.; Oblock, S.T.; Harris, L.R. A review of the environmental fate, effects, and exposures of bisphenol A. *Chemosphere* 1998, *36*, 2149–2173.
41. Sekela, M.; Brewer, R.; Moyle, G.; Tuominen, T. Occurrence of an environmental estrogen (4-nonylphenol) in sewage treatment plant effluent and the aquatic receiving environment. *Water Sci. Technol.* 1999, *39*, 217–220.
42. Ferguson, P.L.; Iden, C.R.; Brownawell, B.J. Distribution and fate of neutral alkylphenol ethoxylate metabolites in a sewage-impacted urban estuary. *Environ. Sci. Technol.* 2001, *35*, 2428–2435.
43. Isobe, T.; Nishiyama, H.; Nakashima, A.; Takada, H. Distribution and behavior of nonylphenol, octylphenol, and nonylphenol monoethoxylate in Tokyo metropolitan area: Their association with aquatic particles and sedimentary distributions. *Environ. Sci. Technol.* 2001, *35*, 1041–1049.
44. Singer, H.; Müller, S.; Tixier, C.; Pillonel, L. Triclosan: Occurrence and fate of a widely used biocide in the aquatic environment: Field measurements in wastewater treatment plants, surface waters, and lake sediments. *Environ. Sci. Technol.* 2002, *36*, 4998–5004.
45. Clara, M.; Strenn, B.; Gans, O.; Martinez, E.; Kreuzinger, N.; Kroiss, H. Removal of selected pharmaceuticals, fragrances and endocrine disrupting compounds in a membrane bioreactor and conventional wastewater treatment plants. *Water Res.* 2005, *39*, 4797–4807.
46. Ying, G.G.; Kookana, R.S.; Dillon, P. Sorption and degradation of selected five endocrine disrupting chemicals in aquifer material. *Water Res.* 2003, *37*, 3785–3791.

47. Li, C.; Li, X.Z.; Graham, N.; Gao, N.Y. The aqueous degradation of bisphenol A and steroid estrogens by ferrate. *Water Res.* 2008, *42*, 109–120.
48. Müller, S.; Schlatter, C. Estrogenic potency of nonylphenol *in vivo*—A case study to evaluate the relevance of human non-occupational exposure. *Pure Appl. Chem.* 1998, *70*, 1847–1853.
49. National Library of Medicine. Triclosan. Available online: http://toxnet.nlm.nih.gov/cgi-bin/sis/search/a?dbs+hsdb:@term+@DOCNO+7194 (accessed on 12 October 2012).
50. Cousins, I.T.; Staples, C.A.; Klečka, G.M.; Mackay, D. A multimedia assessment of the environmental fate of bisphenol A. *Hum. Ecol. Risk Assess.* 2002, *8*, 1107–1135.
51. Nghiem, L.D.; Schäfer, A.I. Adsorption and transport of trace contaminant estrone in NF/RO membranes. *Environ. Eng. Sci.* 2002, *19*, 441–451.
52. Hurwitz, A.R.; Liu, S.T. Determination of aqueous solubility and pKa values of estrogens. *J. Pharm. Sci.* 1977, *66*, 624–627.
53. Ahel, M.; Giger, W. Partitioning of alkylphenols and alkylphenol polyethoxylates between water and organic solvents. *Chemosphere* 1993, *26*, 1471–1478.
54. Machatha, S.G.; Yalkowsky, S.H. Comparison of the octanol/water partition coefficients calculated to experimentally determined values. *Int. J. Pharm.* 2005, *294*, 185–192.
55. Gerstl, Z. Estimation of organic chemical sorption by soils. *J. Contam. Hydrol.* 1990, *6*, 357–375.
56. Patrolecco, L.; Capri, S.; Angelis, S.D.; Pagnotta, R.; Polesello, S.; Valsecchi, S. Partition of nonylphenol and related compounds among different aquatic compartments in Tiber River (Central Italy). *Water Air Soil Pollut.* 2006, *172*, 151–166.
57. Porter, A.J.; Hayden, N.J. Nonylphenol in the Environment: A Critical Review. Available online: http://www.emba.uvm.edu/~nhayden/npreview.Pdf (accessed on 12 August 2002).
58. Huang, W.; Weber, W.J. A distributed reactivity model for sorption by soils and sediments. Relationships between desorption, hysteresis, and the chemical characteristics of organic domains. *Environ. Sci. Technol.* 1997, *31*, 2562–2569.

Chapter 5

Transport of Pathogen Surrogates in Soil Treatment Units: Numerical Modeling

Ivan Morales[1], Janet A. Atoyan[2], José A. Amador[2] and Thomas Boving[1,3]

[1]Department of Civil and Environmental Engineering, University of Rhode Island, Kingston, RI 02881, USA; E-Mail: ivan_morales@my.uri.edu

[2]Laboratory of Soil Ecology and Microbiology, University of Rhode Island, Kingston, RI 02881, USA; E-Mails: jatoyan@mail.uri.edu (J.A.At.); jamador@uri.edu (J.A.Am.)

[3]Department of Geosciences, University of Rhode Island, Kingston, RI 02881, USA

ABSTRACT

Segmented mesocosms (n = 3) packed with sand, sandy loam or clay loam soil were used to determine the effect of soil texture and depth on transport of two septic tank effluent (STE)-borne microbial pathogen surrogates—green fluorescent protein-labeled *E. coli* (GFPE) and MS-2 coliphage—in soil treatment units. HYDRUS 2D/3D software was used to model the transport of these microbes from the infiltrative surface. Mesocosms were spiked with GFPE and MS-2 coliphage at 10^5 cfu/mL STE and 10^5–10^6 pfu/mL STE, respectively. In all soils, removal rates were >99.99% at 25 cm. The transport simulation compared (1) optimization; and (2) trial-and-error modeling approaches. Only slight differences between the transport parameters were observed between these approaches. Treating both the die-off rates and attachment/detachment rates as variables resulted in an overall better model fit, particularly for the tailing phase of the experiments. Independent of the fitting procedure, attachment rates computed by the model were higher in sandy and sandy loam soils than clay, which was attributed to unsaturated flow conditions at lower water content in the coarser-textured soils. Early breakthrough of the bacteria and virus indicated the presence of preferential flow in the system in the structured clay loam soil, resulting in faster movement of water and microbes through the soil relative to a conservative tracer (bromide).

INTRODUCTION

Septic tank effluent (STE) contains pathogenic microorganisms, such as enteric viruses and bacteria, which can cause waterborne diseases and pose a public health risk if not properly treated. Onsite wastewater treatment systems (OWTS) are commonly used in suburban and rural areas in the United States and elsewhere. A conventional OWTS consists of a septic tank and associated soil treatment unit (STU), or drainfield, where attenuation and/or removal of microbial contaminants can take place through interactions with the soil, preventing their transport to groundwater. Nevertheless, contamination of groundwater below the STU is a concern, especially in areas with shallow groundwater tables.

In order to protect drinking water, the separation distance between the infiltrative surface of the STU and fluctuating water table has to be determined on a site-by-site basis, and seasonal variations in separation distance have to be considered to avoid microbial contamination. The US EPA [1] recommends a minimum separation distance of 45 cm, regardless of soil chemical and physical characteristics. However, separation distance requirements in the USA vary widely by state, region and sensitivity of receiving waters to contaminant load. Furthermore, differences in soil properties (e.g., texture, structure, pH) are known to affect STU performance, which may lead to differences in removal of viruses and bacteria [1,2,3].

A number of studies have investigated the removal efficiency of bacteria in STUs and the processes involved. Crites [4] suggested that bacterial removal or inactivation in STUs is associated with predation by bactrivorous organisms and exposure to sunlight. Mechanical filtration, adsorption, and flow rate also have a significant effect on removal of pathogenic bacteria [2,5,6]. All of these processes are influenced by soil texture and structure. Fine textured and poorly structured soils are expected to remove bacteria mainly through mechanical filtration because of the smaller pore sizes and lower hydraulic conductivity of those soils. Together with a larger surface area, this results in higher rates of bacteria adsorption [2]. In contrast, coarse and well-structured soils have larger pores and lower porosity values, which allow for better aeration that promotes microbial predation and attenuation [7].

Viruses are thought to be removed in STUs through adsorption to soil particles rather than by mechanical filtration [8,9]. Viruses have a smaller diameter compared to soil pores, which prevents them from being trapped in the pore space. Adsorption of viruses is a function of the physical and chemical properties of the soil, particularly pH, organic matter content, and water content [10,11,12,13].

Mature OWTS systems develop a biological growth layer of low permeability at the infiltrative surface of the STU, known as a biomat. Typically, the biomat extends up to 2 cm below the water-soil interface [14,15]. It may enhance the inactivation of microbes through mechanical filtration, because partial clogging of smaller soil pores results in reduced infiltration rates and the development of unsaturated flow conditions in the underlying soil profile [14,15]. Unsaturated flow conditions result in longer contact times between microbes and soil particles, which improves the pathogen removal efficiency of the soil treatment unit [15,16].

The retention of microorganisms in soil can be affected by preferential flow, which may be associated with pathways created by plant roots and earthworms, the presence of interaggregate spaces [17,18], and differences in hydraulic conductivity within the soil strata [19]. Preferential flow increases the travel velocity of the aqueous phase, allowing for faster and deeper movement of microbes into the soil profile [20,21,22].

The complex nature of pathogen removal and inactivation in the STU presents a difficult problem with respect to predicting OWTS effectiveness. Contaminant transport models can be used to predict the microbial transport in soils and to help elucidate the factors that control microbial fate as STE moves through the soil profile.

Several models have been developed to simulate virus and bacteria transport in soil. The commercially-available HYDRUS software package is widely used to simulate microbial transport and fate processes, including the transport of viruses, bacteria, and colloids based on either attachment/detachment theory or filtration theory in variably saturated porous media [22,23,24,25,26]. The model supports an interactive graphics-based user interface, and the computational program numerically solves the Richards equation for variably saturated water flow, as well as the advection-dispersion equations for both heat and solute transport. There are HYDRUS versions available with one-, two- and three-dimensional transport modeling capabilities.

The use and calibration of sophisticated transport models, like HYDRUS 2D/3D, permits investigation of the role of microbial inactivation, removal, and transport processes in homogeneous/heterogeneous soil media by quantifying parameters, such as die-off rates in water and soil or attachment/detachment rates [23]. The calibrated transport parameters can be used to calculate microbial removal as a function of distance between the infiltrative surface and the water table, thus permitting comparison among different soils. HYDRUS is a valuable and accepted tool for drinking water protection and water resources management purposes. Because of its many capabilities and multi-dimensional functionality, HYDRUS 2D/3D was chosen for modeling our test data.

The objectives of our research project were to: (1) determine the extent to which removal of two microbial pathogen surrogates—a coliphage virus and a tracer bacterium—is affected by soil texture and depth; (2) measure the survival of the coliphage virus and tracer bacterium in sterile and non-sterile unsaturated soil and STE; and (3) model microbial transport and estimate transport parameters. The results were intended to define and evaluate the potential risk of microbial contamination of groundwater resulting from soil-based treatment of STE. In this paper, we focus on the modeling of microbial transport and how different approaches to modeling—numerical optimization *versus* visual assessment—best describe experimental data.

MATERIAL AND METHODS

Replicate ($n = 3$) segmented mesocosms were constructed to investigate the removal of microbial pathogen surrogates. The mesocosms consisted of straight-sided polypropylene Nalgene jars (10.5 cm height, 6.5 cm diameter) connected to each other with plastic tubing (Figure 1). Sample ports between jars allowed for collecting drainage water directly below the infiltrative surface (4 cm), and at succeeding 10.5 cm depths intervals (14.5 cm, 25 cm, 35.5 cm), hereafter referred to as 4 cm, 14 cm, 25 cm, and 35 cm. The mesocosms were packed with (1) a sandy, B and C horizon soil from Kingston, Rhode Island, USA; (2) a sandy loam soil from Golden, Colorado, USA; or (3) a structured clay loam soil from Griffin, Georgia, USA. These soils are typical of STUs in their respective areas. Select physical and chemical properties of the septic tank influent and for the test soils are shown in Table 1 and Table 2, respectively. The remaining space in the top, 4 cm mesocosm was packed with gravel to simulate a layer of rock that is typically placed at the bottom of conventional OWTS leachfield trench. After initial packing, the mesocosms were saturated by pumping three pore volumes of clean water upward from the bottom of the mesocosms. Afterwards, the water was allowed to drain freely. The mesocosms were maintained in the dark at room temperature (19–21°C).

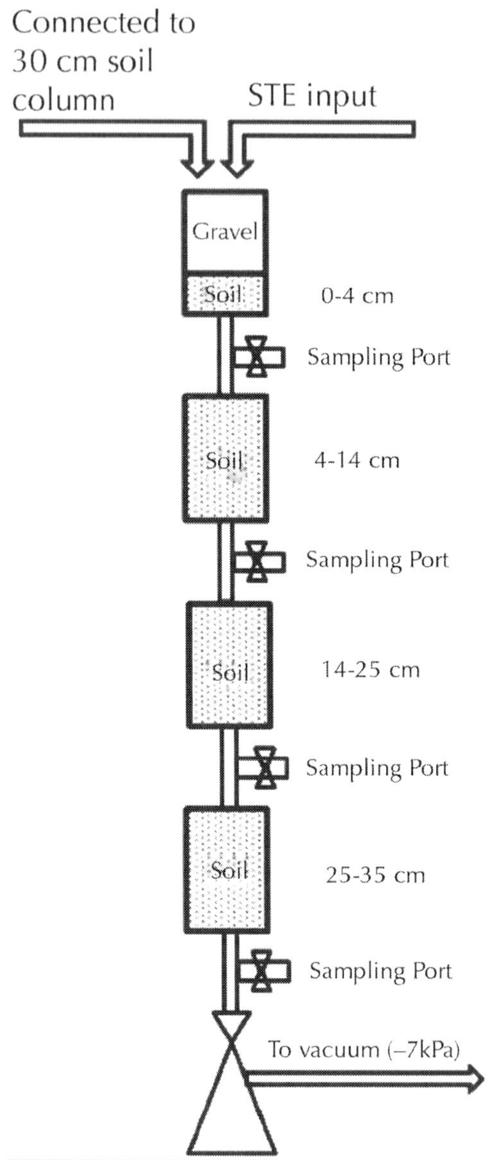

Figure 1: Schematic of experimental setup. Drawing is not to scale.

Septic tank effluent was obtained every 7–10 days from an OWTS serving a group home managed by the Rhode Island Department of Mental Health, Rehabilitation and Hospitals in southern Rhode Island, and stored in the dark at room temperature. Select chemical and microbiological properties of the STE are provided in Table 2. STE was applied to the infiltrative surface of the mesocosms every 12 h at a rate of 2.4 cm/day (0.6 gal./sq.ft./day) using a programmable peristaltic pump (IsmaTec, IDEX Health and Science GmBH, Wertheim, Germany). To mimic the soil atmosphere of a full-scale operating STU trench at the infiltrative surface, the headspace of the top mesocosm (4 cm) was vented to a 30-cm column of soil. Vacuum pressure of –7 kPa was used to approximate the capillary suction from underlying unsaturated soil. Water samples were collected either 3 h to 5 h after dosing (sandy and sandy loam soils) or during dosing (clay loam soil). The samples were withdrawn from the sample ports below the mesocosms using UV-sterilized glass vials and a mild vacuum, with 1–3 mL of water collected from each mesocosm per sampling event.

Table 1: Chemical and microbial properties of septic tank effluent.

Parameter	BOD_5 (mg/L)	pH	Dissolved O_2 (mg/L)	Fecal coliforms (cfu/mL)	Coli-phages (pfu/mL)	Total P (mg/L)	Total N (mg/L)
Average	224	6.82	0	2.93×10^3	0	6	30
Max	383	7.20	0	8.70×10^3	2	8	52
Min	45	6.43	0	1.00×10^2	0	2	0
n	35	36	23	32	32	33	33

Table 2: Soil properties prior to septic tank effluent addition (initial), and after biomat formation

Textural class	**Depth (cm)**	**pH**	**Total C (g/kg soil)**	**Total N (g/kg soil)**	**Water content (g/g soil)**
Sand	Initial	3.2	3.6	0.3	-
	4	6.8	3.5	0.3	0.15
	14	5.9	3.3	0.3	0.09
	25	5.9	2.5	0.1	0.09
	35	8.0	2.7	0.2	0.04

Sandy loam	Initial	6.0	12.9	1.1	-
	4	6.7	9.3	1.0	0.23
	14	6.7	9.9	0.9	0.21
	25	7.0	10.2	0.9	0.21
	35	7.1	10.7	1.0	0.14
Clay loam	Initial	4.2–4.6	2.7–4.8	0.3–0.6	-
	4	6.0	5.4	1.0	0.32
	14	6.1	4.7	0.9	0.32
	25	5.7	2.7	0.8	0.30
	35	5.4	2.6	0.8	0.27

STE was analyzed for dissolved oxygen immediately after collection using the azide modification of the Winkler titration method [27]. The pH was determined using a combination pH electrode and a Model UB-10 pH meter (Denver Instruments, Denver, CO, USA). STE was analyzed for fecal coliform bacteria using the membrane filtration method [27], and for bacteriophage capable of growing on *E. coli* (K12) using the plaque-forming assay of Adams [28]. Five-day biochemical oxygen demand (BOD_5) was determined following standard procedures [27]. Total P and total N were measured in STE using the persulfate digestion method [27], followed by colorimetric analysis [29,30].

Bromide Tracer

Bromide (Br) is a conservative tracer that permits measuring the breakthrough time of the aqueous solution and relates it to the (retarded) transport of either the bacterial or viral tracers. Tracer tests were conducted by spiking the STE influent with KBr (~20 mg Br^-/L). Bromide concentrations were measured using the method of Lepore and Barak [31]. The bromide tracer test data were analyzed with the public domain model CXTFIT to determine the dispersivity (λ) value of each test material [32]. The data were then used for calibration of the transport model.

E. coli Tracer

A novel strain of *E. coli* (BTF 132) (Biomérieux, Hazelwood, MO, USA) was used as a bacterial tracer. The strain has a gene for the production of green fluorescent protein (GFP) inserted into the chromosome. Because the gene is chromosomal and not easily lost, this GFP-labeled *E. coli* strain is ideal for use as a bacterial tracer [33]. Bacterial colonies formed on agar plates glow green under UV light; therefore, it is possible to differentiate between the bacteria that were added to the mesocosms and native fecal coliform bacteria, which do not fluoresce.

For each bacterial addition experiment, GFP *E. coli* were grown overnight at 37 °C in lysogeny broth (LB) and then diluted in phosphate-buffered saline (PBS) solution to ~5 × 10^6 cfu/mL. Approximately 10 mL of the diluted GFP *E. coli* culture was added to each mesocosm over a 37-h period, coincident with the STE dosings (a total of four, 2.5-mL doses). GFP *E. coli* were enumerated using a membrane filtration method with visualization under UV light [27].

Virus Tracer

The bacteriophage MS-2 was used as a tracer. MS-2 is a single-stranded RNA coliphage with a 25-nm diameter and an isoelectric point of 3.9 [13]. *E. coli* strain K12 was used as the host for the bacteriophage. MS-2 bacteriophage was obtained from the Colorado School of Mines (Golden, CO). For each virus addition experiment, MS-2 was diluted in PBS to ~5 × 10^6 pfu/mL and added as described above for the *E. coli* tracer experiment. The bacteriophage in the collected samples were enumerated using the plaque-forming assay of Adams [28] on LB agar plates, which were incubated for ~4 h at 37 °C, followed by incubation at room temperature overnight before counting plaques in the host lawn.

Survival in Soil and STE

Experiments were conducted to determine the survival of the microbial pathogen surrogates in soil and STE. For soil, 2 g (air-dry weight) of soil from each of the three soil types were placed in plastic scintillation vials, in triplicate. Prior to use, the soil was either air-dried or sterilized (121 °C for 60 min on 5 consecutive days). GFP *E. coli* or MS-2 bacteriophage suspension was added to the soil to final concentration of ~2.4 × 10^5 cfu (pfu)/g soil. Three replicate vials were sacrificed periodically to enumerate the pathogen surrogates. Microorganisms were extracted with 20 mL of sterile PBS (for *E. coli*) [34] or sterile 1.5% beef extract, pH 8.7 (for MS-2) [13] added to each vial, after which the vials were placed on a reciprocal shaker for 10 min. GFP *E. coli* and MS-2 bacteriophage were enumerated as described above.

To determine survival in STE, four 250-mL polypropylene bottles containing 100 mL of effluent were amended with: (1) ~4.8×10^2 cfu GFP *E. coli*/mL; (2) ~5.2×10^6 cfu GFP *E. coli*/mL; (3) ~2.3×10^2 pfu MS-2 coliphage/mL; or (4) ~4.4×10^6 pfu MS-2 coliphage/mL. The bottles containing the amended STE were then incubated at room temperature, in the dark. Samples were analyzed for *E. coli* and MS-2 as described above. Initial concentrations were determined by identical dilutions using PBS in place of STE followed by immediate enumeration. This experiment was repeated three times.

Soil Properties

Sand, sandy loam, and clay loam soils were analyzed prior to the start of the experiment and after STE dosing for 27, 31 and 44, weeks, respectively. After 44 weeks, all mesocosms had developed a biomat layer that extended over the entire thickness of the soil (4 cm) at the infiltration surface. The total carbon and nitrogen content of the soil was determined using a Carlo Erba EA1108 CHN analyzer (Lakewood, NJ, USA). The soil pH was determined using a 1:5 soil/water ratio with a combination pH electrode and a Model UB-10 pH meter (Denver Instruments). Particle size analysis was conducted using the pipette method [35]. The water content was determined gravimetrically.

Bacteria and Virus Transport Modeling

HYDRUS 2D/3D was used to simulate the transport of microbes in the segmented mesocosms at different depths. The model simulates virus and bacteria transport and fate processes based on a modified form of the advection-dispersion equation [23] (Equation (1)):

$$\frac{\partial \theta C}{\partial t} + \rho \frac{\partial S_e}{\delta t} + \rho \frac{\partial S_1}{\partial t} = \frac{\delta q_i C}{\delta x_i} + \frac{\delta}{\delta t}\left(\theta D_{ij}^{w} \frac{\delta C}{\delta xj} \right) - \mu_w \theta C - \mu_s \rho \left(S_e + S_1 \right) \quad (1)$$

Where C and S are the (virus, bacteria) solution concentration [NcL^{-3}] and the solid phase (virus, bacteria) concentration [NcM^{-1}], respectively. Subscripts e and 1 represent equilibrium and kinetic sorption sites. *Nc* is a number of microbial particles, and μ_w and μ_s represent inactivation and degradation processes (die-off) in the liquid and solid phases, respectively. D^w_{ij} is the dispersion coefficient for the liquid phase [L^2T^{-1}], θ is the volumetric water content [L^3L^{-3}], ρ is the bulk density of porous medium [ML^{-3}], and q is the specific discharge [L^3T^{-1}]. Mass transfer between the aqueous and solid phases can be described as in (Equation (2)):

$$\rho\frac{\partial S}{\partial t} = \theta K_a C - K_d \rho S \quad (2)$$

Where k_a is the first-order attachment coefficient [T^{-1}] and k_d the first-order detachment coefficient [T^{-1}]. According to Simunek [23], the attachment and detachment coefficients are strongly dependent upon the water content, with attachment significantly increasing as the water content decreases. Linear adsorption kinetics were assumed. The chemical non-equilibrium model was used, with 50% of all sorption sites assumed to sorb instantaneously and the other 50% are governed by kinetic sorption.

A 2D columnar hydraulic model domain was developed (Figure 2) with observation nodes at 4 cm, 14 cm and 25 cm. The model geometry closely resembled the experimental set-up of the *E. coli* and MS-2 coliphage column experiments, that is, the model boundary on top of the flow domain was open to the atmosphere. The pulsed injections of microbial and conservative tracer (bromide) occurred into an initially unsaturated columnar environment. The bottom boundary was set at a fixed negative pressure head −7 kPa to simulate drainage into the unsaturated zone beneath. No-flux boundaries define the system along the column side boundaries. The flow and transport through the soil was modeled either with or without a 4-cm thick biomat layer. The properties of the porous materials were obtained from the HYDRUS soil catalog [36]. Based on literature data, the diameter was set at 1.1 μm for *E. coli* and 0.025 μm for the MS-2 coliphage [37].

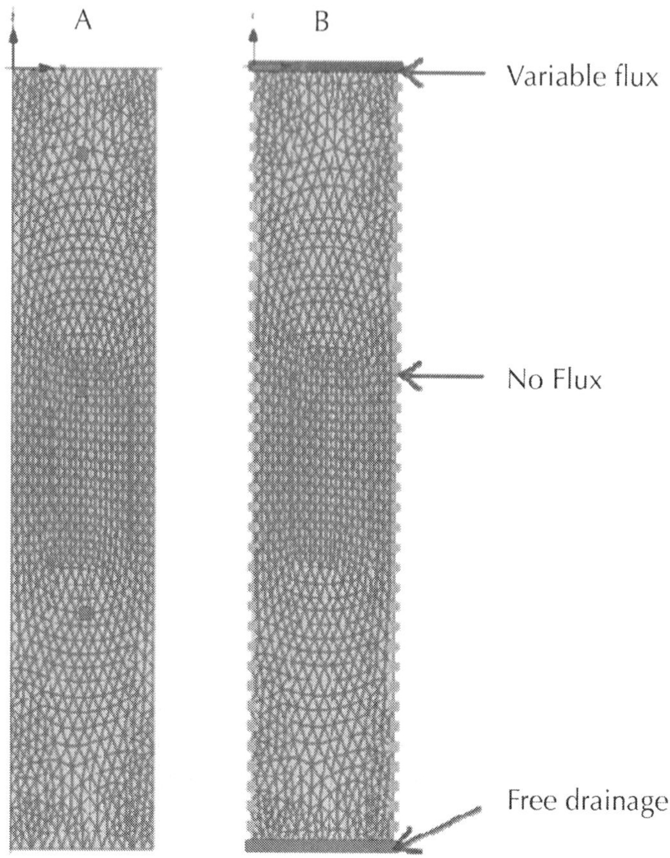

Figure 2: Hydraulic model domain with (A) observation nodes at 4 cm, 14 cm and 25 cm depths; (B) boundary conditions.

RESULTS

The bromide tracer test data and CXTFIT 2.1code were used to determine the column system dispersivity (λ) for all three soils. Model fits were good with R^2 values ranging from 0.97 to 0.99. The dispersivity value calculated by CXTFIT 2.1 was approximately 0.289 cm, which is typical for these types of experiments, and is consistent with the range of values (0.06 to 0.816 cm) reported by others [38,39,40,41,42,43]. Next, the hydraulics of the HYDRUS model domain was calibrated using the conservative tracer breakthrough curves (BTC). The tracer test results were fitted for each of the three soil column depth intervals (0–4 cm, 4–10.5 cm, and 10.5–25 cm). The data obtained at the 35 cm sample port was not fitted because *E. coli* and MS-2 phage concentrations were always below detection limit at that depth. The model results were plotted against the observed data (Figure 3).

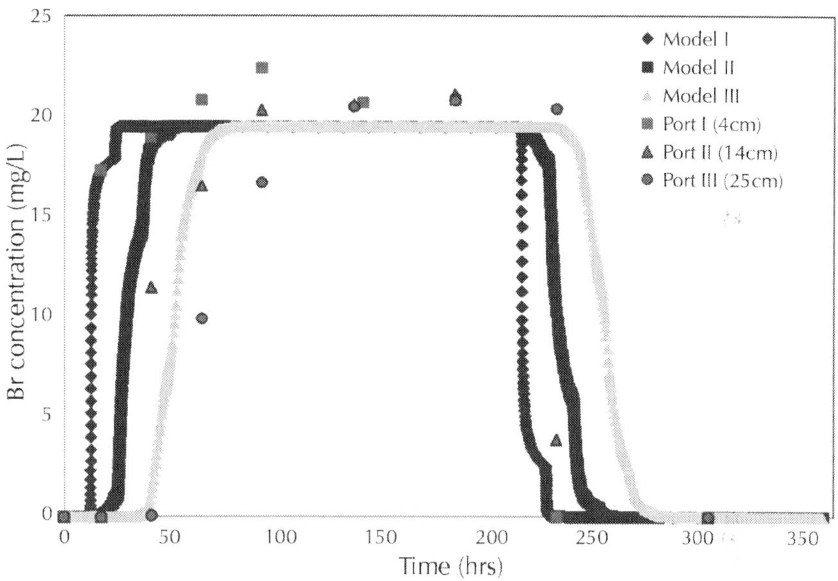

Figure 3: Bromide tracer test results and best fit obtained for the three soil column depth intervals (0–4 cm, 4–14 cm, and 14–25 cm) in sandy soil.

The experimental bacterial transport data were fitted to HYDRUS utilizing the model's attachment/detachment module. The data were fitted in two steps: (1) inverse solution, keeping constant the STE and soil die-off rates values (Table 3 and Table 4), to determine the optimized attachment/detachment rates; and (2) a trial-and-error process in which die-off and attachment/detachments rates were modified simultaneously until an acceptable graphical fit was achieved. During the trial-and-error process, the emphasis was on achieving the best fit of the tailing end of experimental data. The best-fit simulations of the bacteria and virus test data from mesocosms are shown in Figure 4 and Figure 5.

Table 3: Modeled green fluorescent protein (GFP) *E. coli* and MS-2 phage peak concentrations for sandy, sandy loam, and clay loam soil using optimization or trial-and-error approaches. Values in parentheses are percent removal rates. The microbial tracers (C_0 = initial concentration) were applied together with STE at a hydraulic loading rate of 2.4 cm/day every 12 h over 37 h. BD: below detection limit

Micro-organism	Soil type	C_0 (cfu/mL)	Simulation type	Peak concentration (cfu/mL)		
				4 cm	14 cm	25 cm
E. coli	Sandy	1.0×10^5	Trial/Error	5.42×10^4 (45.80)	8.76×10^2 (99.12)	1.35×10 (99.99)
			Optimized	5.46×10^4 (45.40)	9.32×10^2 (99.07)	1.53×10 (99.98)
	Sandy loam	5.9×10^6	Trial/Error	2.45×10^6 (58.47)	5.30×10^4 (99.10)	1.46×10^3 (99.98)
			Optimized	2.60×10^6 (55.93)	7.64×10^4 (98.71)	2.74×10^3 (99.95)
	Clay loam	5.9×10^6	Trial/Error	1.02×10^6 (82.71)	8.08×10^2 (99.99)	BD (100)
			Optimized	9.36×10^5 (84.14)	4.62×10^2 (99.99)	BD (100)
MS-2 Phage	Sandy	1.0×10^5	Trial/Error	1.85×10^3 (98.15)	BD (100)	BD (100)
			Optimized	2.38×10^4 (76.20)	BD (100)	BD (100)
	Sandy loam	5.9×10^6	Trial/Error	BD	BD	BD
			Optimized	BD	BD	BD
	Clay loam	5.9×10^6	Trial/Error	BD	BD	BD
			Optimized	BD	BD	BD

Table 4: Measured GFP *E. coli* die-off rates in soil and septic tank effluent

Medium	Die-off rate (h^{-1})	
	Measured	R^2
Sand	0.0617	0.791
Sandy loam	0.0298	0.965
Clay loam	0.2476	0.965
Septic tank effluent	0.0824	0.891

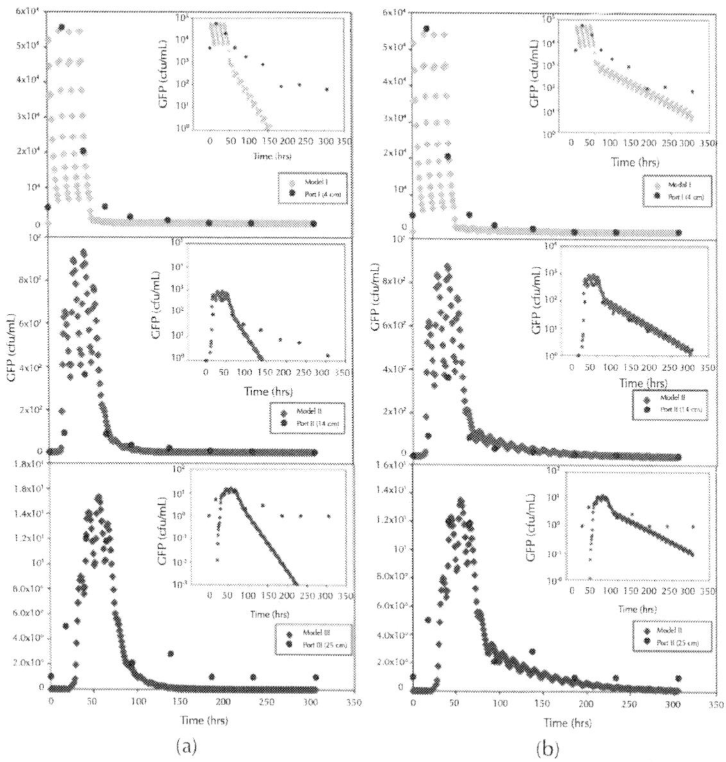

Figure 4: Experimental GFP *E. coli* data and HYDRUS optimization (**a**) and trial-and-error; (**b**) simulation for the sandy soil at 4 cm, 14 cm and 25 cm depth. The log-normal inserts emphasize the tailing phase. 4441/6/4/818/ htm#sthash.y5kl8AcD.dpuf.

for $Sink_S$ for the clay loam soil, which varied by about an order of magnitude. Overall, the trial-and-error die-off rates tended to be lower than the measured values used in the optimization procedure. Lower trial-and-error die-off rates appeared to have been compensated for by attachment rates that were approximately 2 to 3 times greater than those obtained by optimization. In the case of the sandy soil attachment rate, the results from both estimation methods resulted in identical outcomes. By contrast, detachment rates were 37 to 74 times lower than attachment rates for trial-and-error and 1.3 to 37 times lower for the optimized simulation. The optimized detachment rate in clay loam soil indicates quasi-irreversible (1×10^{-7} h^{-1}) detachment of GFP *E. coli*. Overall, the combination of lower die-off rates and greater attachment rates resulted in a better capture of the tailing phase when using the trial-and-error procedure (log-log inserts in Figure 5 through Figure 7).

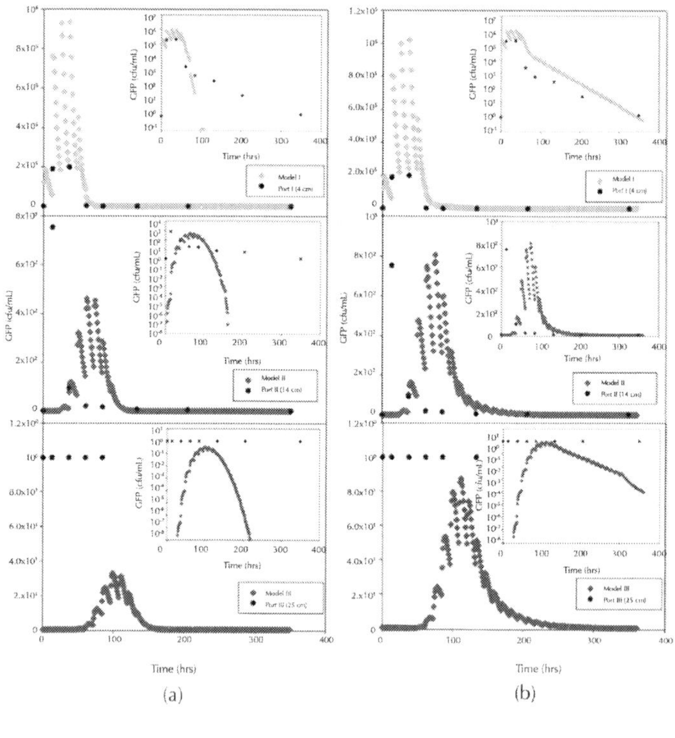

Figure 6: Experimental GFP *E. coli* data and HYDRUS optimization (a) and trial-and-error (b) simulation for the clay loam soil at 4 cm, 14 cm and 25 cm depth. The log-normal inserts emphasize the tailing phase.

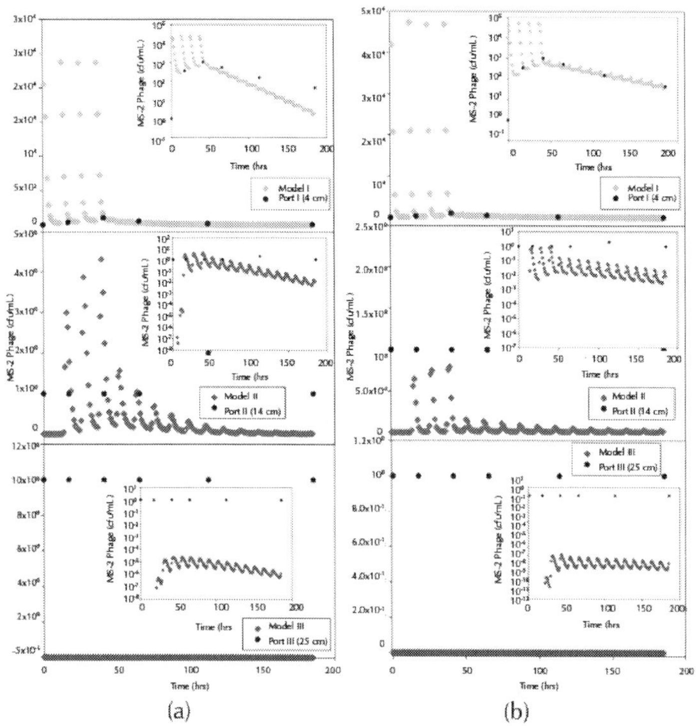

Figure 7: Experimental MS-2 phage data and HYDRUS optimization (a) and trial-and-error; (b) simulation for the sandy soil at 4 cm, 14 cm and 25 cm depth. The log-normal inserts emphasize the tailing phase.

Only the sandy soil experiment produced sufficient breakthrough data to attempt a simulation of the virus data. The liquid phase die-off rate was about half of the solid phase die-off rate for the optimization procedure ($R^2 = 0.76$), but more than an order of magnitude greater for the trial-and-error simulation. On the other hand, the trial-and-error and optimization virus attachment and detachment rates were very similar to each other. Overall, the results indicate that virus attachment rates were more than an order of magnitude higher than those for bacteria, while bacteria and virus detachment rates were similar.

Table 5: *E. coli* and virus transport parameters derived from HYDRUS 2D/3D [23]. Non-detects of virus tracer in sandy loam and clay soils prevented modeling of these experiments

Micro-organism	Soil type	Simulation type	Transport parameters				R^2
			$Sink_L$ (h^{-1})	$Sink_S$ (h^{-1})	Attach (h^{-1})	Detach (h^{-1})	
E. coli	Sand	Optimization	0.0824	0.062	0.163	0.0044	0.91
		Trial/Error	0.085	0.019	0.163	0.0023	
	Sandy loam	Optimization	0.0824	0.0298	0.026	0.0199	0.83
		Trial/Error	0.0298	0.020	0.085	0.0023	
	Clay loam	Optimization	0.0824	0.2476	0.078	0.0000001	0.99
		Trial/Error	0.020	0.036	0.133	0.0018	
***MS-2* Phage**	Sand	Optimization	0.0271	0.0425	1.00	0.0064	0.76
		Trial/Error	0.750	0.017	0.82	0.0040	

Notes: $Sink_L$, aqueous phase die-off rate (STE); $Sink_S$, solid phase die-off rate; Attach, attachment rate; Detach, detachment rate.

DISCUSSION

The bacteria die-off rates measured for the three soil types were different (Table 5), which provides evidence for the effect of local environmental soil conditions on bacteria die-off rates [44]. Chao and Feng [45] studied the survival of *E. coli* HB101 strains added to a silt loam soil at 30 °C, resulting in die-off rates ranging from 0.04 d^{-1} to 0.20 d^{-1} 4(0.0017 h^{-1} to 0.0083 h^{-1}). Powelson and Mills [46] reported *E. coli* die-off rates of 0.0259 h^{-1} and 0.0693 h^{-1} in sand columns under saturated and unsaturated conditions, respectively. *E. coli* isolated from STE collected from an OWTS near Lake Okareka, New Zealand, were investigated to elucidate microbial attenuation and transport through pumice sand aquifers [47]. The results of that study showed soil-attached *E. coli* die-off rates ranged from 2.59 h^{-1} to 4.47 h^{-1}. These studies suggest that solid phase bacteria die-off rates have to be determined under environmental conditions representative of the location where the OWTS system will be built.

The measured die-off rates reported here for all three soil types may be different from *in situ* rates where the soils were collected (Colorado, Georgia, Rhode Island). It is also likely that the liquid phase die-off rates differ among locations because of differences in the chemical, physical and biological properties of wastewater. For the trial-and-error simulations, both the solid and liquid phase die-off rates were treated as variables, whereas they were fixed to the measured values during the optimization procedure (Table 5). The attachment/detachment rates were fitting variables in both procedures. Based on the assumption of location-specific solid and liquid phase die-off rates, treating these rates as variables may be considered for the optimization procedure. A better fit could be obtained by treating the die-off rates as variables, particularly during the tailing phase of each experiment. Further research is needed to confirm this approach.

Average attachment rates, derived from either optimization or trial-and-error procedures, were highest for the sandy soil (0.163 h^{-1} for *E. coli*, 0.91 h^{-1} for MS-2). This result was unexpected because higher attachment rates are typically reported for fine-grained clay materials, rather than sandy soils. In general, the intrinsic lower surface area of coarser soils should result in less adsorption of microbes compared to finer textured soils [10,48,49]. In addition, the smaller pores that are prevalent in fine-grained soils are more effective for mechanical filtration (straining) of microbes than those in coarser porous soils. Conversely, unsaturated soils tend to retain more microbes than saturated soils. That is, with decreasing water content, higher retention of bacteria and viruses in the soil has been observed [24,38,50,51]. Because the air-water interface increases at decreasing water content, the removal and retention of microbes in fine-grained soil, such as the clay loam, should be, at a given water content, greater than in granular soils [24,38,52]. In our study, the water content of the sand and sandy loam soils at the end of the experiment was lower (0.15 and 0.23 g/g, respectively), compared to the clay loam soil (0.32 g/g; Table 2). Therefore, the higher air-water interface in the coarser soils could explain the higher attachment rates, since more water-free surface area is available to interact with the microorganisms. Measurements of the air-water interface area at different saturations in various soil materials would be necessary to confirm this proposition.

The effects of soil texture on microbial removal are expected to be different for bacteria and viruses. In our experiments, MS-2 phages were removed much more effectively than bacteria, sandy loam and clay loam soils removed phages more extensively than sandy soil did. Two main mechanisms have been considered for pathogen removal in soil: (i) mechanical filtration; and (ii) adsorption. For instance, Powelson *et al.* [13] investigated the fate and transport of a *Salmonella* phage in structured soils and found a reduction in virus concentration of about 60% to 90% in clay, clay loam and silt loam soils. In a review of the literature, Amador *et al.* [53] concluded that, although coarser textured soils tend to remove fewer bacterial pathogens than finer textured soils, the depth of treatment is important in order to obtain acceptable removal rates (close to 100%). The authors suggest that, because preferential pathways are more common in fine-grained, structured soils, these pathways facilitate the transport of microbes to greater depths relative to unstructured fine textured soils. In addition, they suggest that the soil texture and depth of soil treatment are not well-correlated variables in virus removal, which is consistent with the hypothesis that virus removal occurs by adsorption processes rather than mechanical filtration. Virus removal by adsorption processes is in agreement with our results, where the model computed higher attachment rates for viruses than bacteria in sandy soil.

On average, the detachment rate for both bacteria and viruses in all soils was 1.6% of the attachment rate (Table 5). The lowest detachment rate values were observed in the structured clay soil, which suggest that bacteria and virus attachment in those soils is practically irreversible. Under those conditions, detachment can be considered negligible. This is consistent with previous studies, which concluded that the attachment of microbes to soil particles is an irreversible process [24,54,55,56].

CONCLUSIONS

Modeling results showed only small differences between attenuation parameters (microbial attachment and detachment rates) obtained by optimization and trial-and-error simulation processes, *i.e.*, results were generally within a factor of three of each other. The microbe detachment rates were about two orders of magnitude lower than the corresponding attachment rates. Low or negligible detachment rates suggest quasi-irreversible adsorption of microbes to soil. GFP *E. coli* concentrations were generally underestimated by the optimization simulation, whereas a better fit was achieved by the trial-and-error procedure, particularly for the tailing phase of each experiment. In case of the liquid and solid phase GFP *E. coli* die-off rates, the results of the optimization and trial-and-error simulations were generally within a factor of three of each other. Overall, the combination of lower die-off rates and higher attachment rates resulted in a better description of the tailing phase when using the trial-and-error procedure.

In general, the fit obtained in the optimization process should improve when the concentration of bacteria or virus is measured more frequently. In addition, the results of the *E. coli* and MS-2 phage die-off rate experiments support the findings of Foppen and Schijven [44] that these measurements should be ideally collected under *in situ* conditions of the sample location, rather than under standard laboratory conditions. This change in procedure would contribute to a better understanding of the effects of the local conditions on the soils and the resulting degradation/attenuation of those microbes.

The experimental data for the structured clay loam suggests that early breakthrough of the bacteria occurred. Although the presence of preferential flow pathways in the mesocosms likely influenced the results, it is not possible to simulate those conditions with existing

54. Rajagopalan, R.; Tien, C. Trajectory Analysis of Deep-Bed Filtration with the Sphere-in-Cell Porous Media Model. *AIChE J.* 1976, *22*, 523–533.

55. Pieper, P.J.; Ryan, N.; Harvey, R.W.; Amy, G.L.; Illangasekare, T.H.; Metge, D.W. Transport and Recovery of Bacteriophage PRD1 in a Sand and Gravel Aquifer: Effect of Sewage-Derived Organic Matter. *Environ. Sci. Technol.* 1997, *31*, 1163–1170.

56. Jarvis, N.J. A Review of Non-Equilibrium Water Flow and Solute Transport in Soil Macropores: Principles, Controlling Factors and Consequences for Water Quality. *Eur. J. Soil Sci.* 2007, *58*, 523–546.

57. Šejna, M.; van Genuchten, M.T.; Šim nek, J. Development and Applications of the HYDRUS and STANMOD Software Packages and Related Codes. *Vadose Zone J.* 2008, *7*, 587–600.

Chapter 6

Spatial Distribution of Fecal Indicator Bacteria in Groundwater Beneath Two Large On-Site Wastewater Treatment Systems

Charles Humphrey[1], Michael O'Driscoll[2] and Jonathan Harris[1]

[1]Environmental Health Sciences Program, East Carolina University, 3400 suite Carol Belk Building, Greenville, NC 27858, USA; E-Mail: Harrisjon@ecu.edu

[2]Department of Geological Sciences, East Carolina University, 204 Graham, Greenville, NC 27858, USA;

ABSTRACT

On-site wastewater treatment systems (OWS) are a common means of wastewater treatment in coastal North Carolina, where the soils are sandy and groundwater is relatively close to the surface (<5 m). Wastewater contains elevated concentrations of pathogenic microorganisms that can contaminate groundwater and surface water if OWS are not operating efficiently and distributing wastewater equally to all drainfield trenches. The objectives of this study were to compare the distribution of fecal indicator bacteria (FIB) in groundwater beneath a large low-pressure pipe (LPP) OWS and a large pump to distribution box system, and to determine the effectiveness of the systems in reducing FIB including total coliform, *E. coli*, and enterococci. Monitoring wells were installed at the fronts and ends of the drainfields for sample collection. Groundwater beneath the LPP had a more homogeneous spatial distribution of *E. coli* and enterococci concentrations and the specific conductivity of groundwater was also more uniform relative to groundwater beneath the distribution box system. Both systems were effective (>99%) at reducing FIB concentrations before discharge to groundwater. Results indicate that the LPP did enhance the distribution of FIB in groundwater beneath the drainfield area relative to the pump to distribution box system. Although the LPP system had a vadose zone over 2 m thinner than the pump to distribution box system, FIB treatment was similar. Enterococci was the most resilient FIB of the three tested.

INTRODUCTION

On-site wastewater treatment systems (OWS) are a common method of wastewater treatment in many countries including the United States, Australia, Canada, New Zealand, and Ireland [1,2,3,4,5]. OWS treat wastewater that contains high concentrations of pathogenic microorganisms such as bacteria, viruses, and protozoa [1,6]. Most OWS include a septic tank, effluent distribution device, drainfield trenches, and soil beneath the trenches. The septic tank provides primary treatment via sedimentation and anaerobic digestion of organic matter. The effluent distribution device distributes wastewater to the drainfield trenches, where effluent is stored until infiltrating the soil. The soil beneath the drainfield trenches provides most of the physical, chemical and biological treatment of wastewater. If OWS do not reduce the concentration of pathogens in wastewater effluent, then groundwater and surface water quality may be degraded and public health may be compromised. For example, studies have shown that concentrations of fecal indicator bacteria (FIB) including *E. coli*, enterococci, total coliform, and/or fecal coliform can be elevated in groundwater adjacent to OWS if there is not sufficient vertical separation between the OWS trenches and water table [7,8,9,10]. Other research has shown that surface water FIB concentrations were elevated in areas adjacent to high densities of OWS [11] and when groundwater levels beneath OWS were elevated surface water FIB concentrations were also elevated [12]. High densities of OWS have been associated with endemic diarrheal illness, with higher OWS densities correlating to higher incidences of bacterial and viral diarrhea in a study in central Wisconsin [13].

Other factors associated with OWS wastewater treatment performance include soil type, loading rates, and system type. OWS installed in soils with higher percentages of fine-textured materials such as silt and clay, typically are more efficient at reducing microbial concentrations than OWS installed in sandy soils, due to the increased reactive surface area and residence time [7,14,15]. Research has also suggested that higher wastewater loading rates may lead to decreased treatment [8,16,17,18] and that OWS that utilize low pressure pipe (LPP) distribution can more efficiently treat wastewater constituents such as bacteria relative to gravity flow effluent distribution systems [16,19,20]. LPP systems include a septic tank and pump tank, a pump that delivers effluent to the drainfield trenches via a pressure manifold connected to small diameter (2.54 to 5.08 cm) PVC conveyance pipes (laterals) that extend the length of each trench (Figure 1). The laterals for each trench are typically the same diameter, and have small holes (0.31 to 0.39 cm) drilled along the lateral about every 1.5 m (in sandy soils) to deliver wastewater to the trench. The ends of the laterals include a "turn-up" that is capped, so that when the pump is activated, the manifold and then all laterals should pressurize and thus deliver equal distribution of effluent across the drainfield. This is in contrast to OWS that use distribution boxes. Effluent distribution with distribution boxes is facilitated by gravitational forces, not pressure, even if the OWS uses a pump. The distribution box is installed at a higher elevation relative to the trenches. The distribution boxes are connected to the drainfield trenches with conveyance pipes that have larger diameter weep holes (1.3 to 1.9 cm) that are more closely spaced (10 cm apart) together than LPP systems [21] (Figure 1). The conveyance pipes are not capped at the end, and thus often deliver more wastewater to the front of the trenches than an LPP system. Because some of the earlier studies suggested that LPP systems could perform better than conventional OWS [16,19,20], there are less stringent siting requirements for LPP systems relative to conventional systems with distribution boxes in North Carolina. For example, in sandy areas, LPP systems can be installed in soils with 15 cm less soil depth and separation to groundwater than conventional OWS [21], and LPP systems typically require about 25%–30% less overall drainfield area than conventional trench systems with distribution boxes [21]. If LPP systems are more effective at distributing wastewater across the entire drainfield area than gravity flow distribution systems, then groundwater characteristics (as influenced by infiltrating wastewater) beneath the LPP systems may also be more homogeneous. However, there is a lack of field research that compares FIB groundwater quality uniformity beneath OWS that use LPP and gravity flow distribution. If the LPP systems are not more effective at distributing effluent across the entire drainfield area than OWS that use distribution boxes, then the overall treatment of wastewater constituents such as pathogenic microorganisms may not be as effective as prior work has suggested.

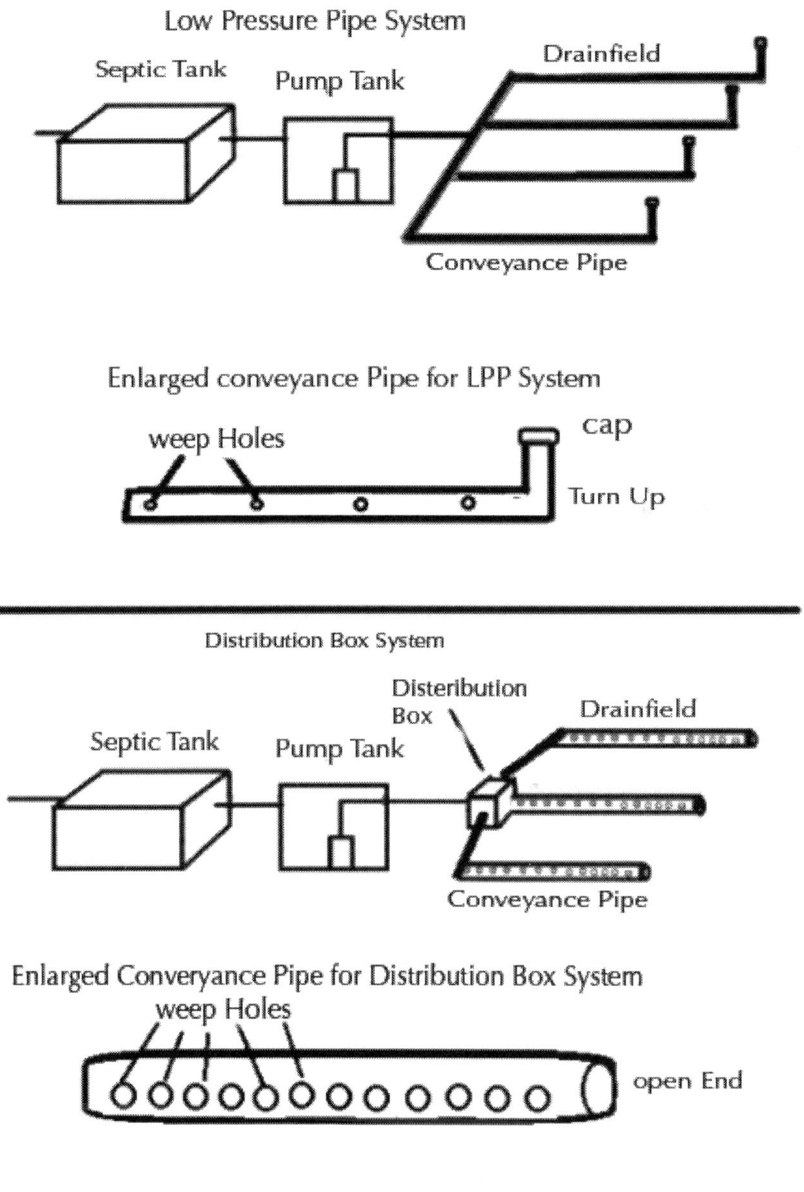

Figure 1: Low pressure pipe (LPP) system (top) and pump to distribution box system (bottom) showing the differences in pipe diameter, weep hole size and spacing, and end of the pipes (capped for LPP, open for distribution box).

The objectives of this research were to: (1) compare the spatial variability of groundwater FIB concentrations beneath OWS that use LPP and distribution boxes for effluent distribution; and (2) evaluate the efficiency of the LPP and pump to distribution box OWS in reducing FIB concentrations.

MATERIALS AND METHODS

Site Selection

Sites were selected at two schools in Craven County North Carolina, USA including James W. Smith Elementary (JWS) and West Craven High School (WCH). The sites were chosen because they were located in coastal North Carolina where issues regarding FIB contamination of surface waters have been documented [8,11]. Both schools have OWS that include large capacity septic tanks, pump tanks with effluent pumps and dual alternatingly dosed disposal fields (Table 1). The OWS at JWS and WCH were routinely maintained by a certified operator with oversight from the Craven County Environmental Health Department. Both systems were in operation during the study and each had 32 drainfield trenches. The OWS at WCH was a replacement system that was installed in 1997. The OWS at JWS was the original system that has been in use since 1987. The OWS at JWS uses a pump to distribution box system and WCH uses a low pressure pipe (LPP) system. Other system characteristics are listed in Table 1.

The soil survey for Craven County listed the soil series as Autryville loamy sand (Loamy siliceous, thermic, Arenic Paleudult) at JWS, and Tarboro sand (Mixed, thermic, typic Udipsamments) at WCH [22]. The soil series at the sites were confirmed by a NC Licensed Soil Scientist working on the project. Soil cores were also collected to depths of 5 m at each site within the drainfield area for descriptive analysis. Grain-size analysis was completed on the soil samples using the GRADISTAT 4.0 method [23]. The mean % sand from soil samples collected from the cores was >90% for both sites.

Table 1: Site and wastewater system characteristics

Site	Install Date	Septic Tank Capacity (L)	Max Design Flow (L/d)	Grease Trap (L)	Pump Tank (L)	Distribution Device	Dispersal Area (m^2)	Vertical Separation (m)	USDA Soil Series
JWS	1987	37,800	37,800	3780	18,144	D-box (2)	892	>4 m	Autryville
WCH	1997	73,827	73,827	11,340	11,340	LPP (2)	1115	>1 m	Tarboro

Groundwater Monitoring Network

Monitoring wells constructed with 3.2 or 5 cm diameter, solid PVC pipe coupled to 0.90 m of well screen were installed near (within 3 m) the corners of both drainfields at each site (~25 m apart) (Figure 2 and Figure 3). Therefore, there were a total of four "front" wells (closest to distribution box/mainfold) and four "end" wells (farthest from the distribution box/manifold). Soil augers and a geoprobe were used to create bore holes to depths below the water table (range: 1 m to 10.5 m) for well placement. Once the wells were placed and driven in the boreholes, the annular space adjacent to the screen was filled with sand, and the annular space above the screen was filled with a mixture of native sand and bentonite. The wells were capped, labeled, and enclosed in valve boxes. The relative elevation of the monitoring wells was determined using a laser level, and the coordinates of the wells were determined using a *Trimble* 6000 *series* GPS (Trimble Navigation Limited, Westminster, CO, USA). Site maps with the monitoring well locations overlain on aerial photographs were created for both sites. Depth to groundwater was determined using a *Solinst* TLC meter (Solinst, Ontario, Canada), and the relative elevation of the groundwater was labeled on the maps. Groundwater flow direction was determined using the groundwater data, and the 3-pt contouring method [24]. Monitoring wells were installed up-gradient and down-gradient from the systems at both sites, after the groundwater flow direction was determined. A spring and stream and were also sampled at JWS down-gradient from the OWS.

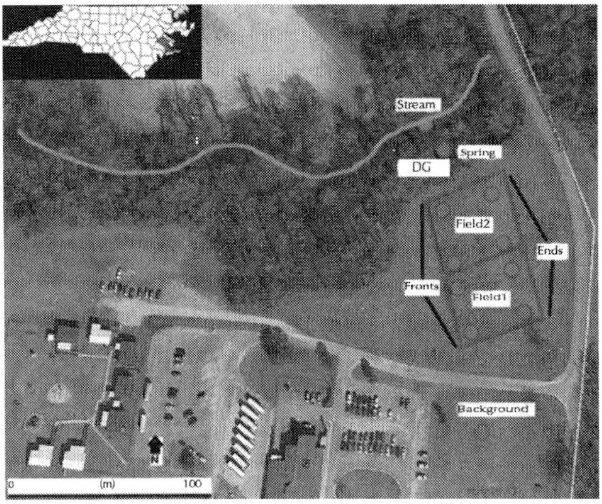

Figure 2: James Smith Elementary School (JWS) in Craven County, North Carolina. The blue circles show monitoring wells, (DG) is down-gradient wells, the red boxes show the boundaries of drainfields 1 and 2. The fronts of the drainfields are located on the west side of the boxes, while the ends of the drainfields are located on the east side of the boxes. Groundwater flow direction is to the north.

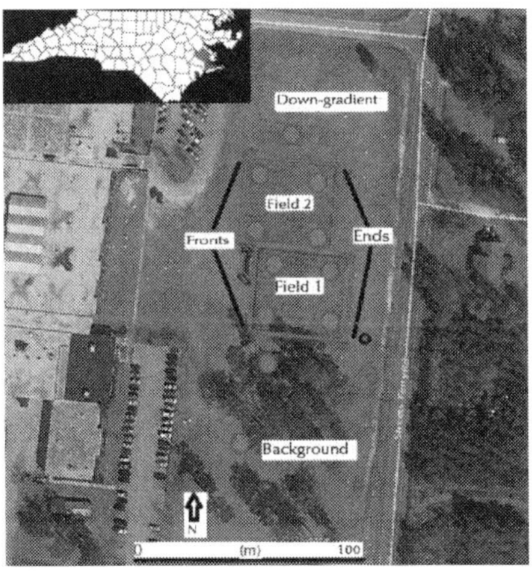

Figure 3: West Craven High School (WCH) in Craven County, North Carolina. The blue circles show the location of monitoring wells, the red boxes show drainfields 1 and 2. The fronts of the drainfields are located on west side of the boxes, while the ends of the drainfields are located on the east side of the boxes. Groundwater flow direction is to the north.

Groundwater Characterization and Analysis

Environmental readings and water samples were collected bi-monthly (4 times) during the study period (November 2012 to May 2013) from each well, septic tank, and adjacent surface water. Samples were collected during periods without rainfall to help reduce the potential for sample contamination. The depth to groundwater at each well was determined using a *Solinst* TLC meter (Solinst, Ontario, Canada). The well was then purged using a sterile (unwrapped immediately prior to use) bailer, allowed to recharge, and a sample was transferred from the bailer to sterile sample bottle. New, neoprene gloves were used for each sample location. Sample bottles were labeled and placed in ice filled coolers for later transport to the East Carolina University Environmental Health Sciences Water Laboratory for analysis. Samples were prepared and incubated within 6 h of collection. Samples from the bailers were also transferred to the multi-meter sensor cup. An YSI 556 multi-meter (YSI Inc., Yellow Springs, OH, USA) was used to determine sample pH, specific conductivity (SC), and temperature in the field. The meter was calibrated prior to each field day. Septic effluent samples were obtained from the distribution box at JWS and from the septic tank at WCH by removing the manhole covers and lowering bailers into the box and tank for sample collection. Effluent samples were also transferred to the sensor cups for environmental readings. Water samples were analyzed for total coliform and *E. coli* using the IDEXX *Colilert* method, and for Enterococci using IDEXX *Enterolert* method (IDEXX Laboratories, Westbrook, Maine). Total coliform and *E. coli* samples were incubated at 35 °C for 24 h and enterococcus samples were incubated at 41 °C for 24 h and then read. Septic effluent samples were diluted (dilution factors of 10–2000×) because of elevated FIB concentrations. For quality control and assurance, field blanks were inserted for approximately 10% of samples.

Spatial Distribution of Fecal Indicators

Groundwater FIB concentrations near the front of each drainfield were compared to groundwater FIB concentrations near the end of each drainfield to determine if there were significant differences in distribution of wastewater and FIB along the length of the drainfield trenches. The data were pooled for each field (Field 1 fronts in comparison to Field 1 ends) and between fields (Field 1 and Field 2 fronts in comparison to Field 1 and Field 2 ends). Groundwater FIB concentrations beneath Field 1 and Field 2 were compared at each site to determine if significant differences were observed between the fields. Mann Whitney tests were performed with *Minitab16* statistical software to determine if differences in concentrations were statistically significant. Groundwater physical and chemical properties including pH, specific conductivity (SC), depth from surface, and temperature were also summarized for Fields 1 and 2, and for proximal and distal ends of the drainfields. FIB data were log transformed and a *Pearson Correlation* test was used to determine if there were significant correlations between the different FIB concentrations, and between FIB concentrations and SC. Correlations between FIB and SC were performed because SC is often elevated in fresh groundwater influenced by effluent [7,25,26].

Treatment Efficiency

FIB concentrations in septic effluent and groundwater beneath the drainfield with the highest FIB concentrations (Field 1 or Field 2 for each site) were compared to determine the effectiveness of the OWS in reducing FIB concentrations in the vadose zone. Septic effluent concentrations of FIB were also compared to groundwater down-gradient and within the flow path of the OWS to determine the effects on groundwater quality down-gradient from the OWS. Groundwater FIB concentrations beneath the drainfield trenches and within the flow paths of the OWS were compared to up-gradient groundwater FIB concentrations to determine if the OWS were influencing groundwater quality. Surface water was also sampled at JWS from a spring located down-gradient and within the groundwater flow path of the OWS, and from an adjacent stream that received the spring discharge. A piezometer adjacent to the spring at JWS was also sampled. Groundwater and surface water FIB concentrations were compared to surface waters standards for the specific indicators. The OWS treatment efficiencies of JWS and WCH were compared. The geometric mean FIB concentrations we observed were not compared to the geometric mean FIB water quality standards for FIB because the sampling frequency used in this study did not conform to the frequency recommended by the US

United States Environmental Protection Agency (EPA). The geometric mean FIB concentrations were instead compared to the single sample FIB standard for infrequently used surface waters suggested by the US EPA.

RESULTS

Groundwater Enterococci Distribution

The geometric mean groundwater enterococci concentrations at JWS (MPN/100 mL) were higher towards the end of the drainfield trenches relative to the front of the trenches for Field 1 (Front: 336, End: 1329) and Field 2 (Front: 36; End:149), and for the pooled data (All Fronts: 110, All Ends: 445) (Figure 4). Statistically significant differences in groundwater enterococci concentrations were observed between the end and front of the trenches for Field 1 ($p = 0.05$) and for the pooled (end *vs.* front) data ($p = 0.004$). Groundwater enterococci concentrations were significantly ($p = 0.0035$) higher beneath Field 1 (667 MPN/100 mL) relative to Field 2 (73 MPN/100 mL). The groundwater geometric mean enterococci concentrations at WCH were elevated toward the end of the trenches beneath Field 1 (End: 1196 MPN/100 mL; Front: 562 MPN/100 mL) and lower toward the end of the trenches beneath Field 2 (Front: 123 MPN/100 mL; End: 92 MPN/100 mL), but the differences were not statistically significant ($p > 0.10$) (Figure 4). When pooling the data, groundwater towards the end of drainfields (geometric mean enterococci concentrations: 331 MPN/100 mL) was not significantly different than groundwater towards the front of the trenches (geometric mean: 262 MPN/100 mL). Groundwater enterococci concentrations were significantly ($p = 0.001$) higher beneath Field 1 (820 MPN/100 mL) relative to Field 2 (106 MPN/100 mL). All field blanks ($n = 12$) were negative for FIB.

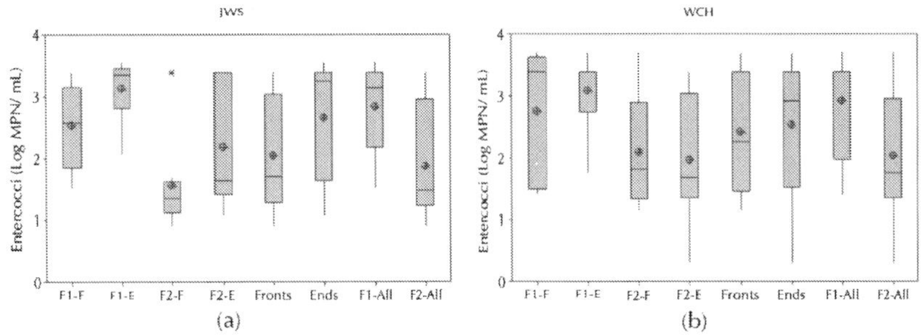

Figure 4: Enterococci concentrations (Log MPN/100 mL) in groundwater beneath the drainfields at James W. Smith (a) and West Craven High (b). Field 1 is F1, Field 2 is F2, front of the field is F, end of the field is E. Data indicate that groundwater enterococci concentrations were significantly higher near the ends of the trenches relative to fronts of the trenches at James Smith, and beneath Field 1 relative to Field 2 at James Smith and West Craven.

Groundwater Total Coliform Distribution

The geometric mean total coliform concentrations in groundwater towards the end of Field 2 (1341 MPN/100 ML) was similar to groundwater coliform concentrations near the front (1034 MPN/100 mL), and for the pooled data (Fronts: 1507, Ends: 1568) (Figure 5). The groundwater total coliform concentrations were elevated beneath Field 1 (2007 MPN/100 mL) in comparison to Field 2 (1178 MPN/100 mL) and the differences were statistically significant (at $p < 0.10$). There was a weak, but significant correlation between enterococci and total coliform concentrations at JWS ($r = 0.446$, $p = 0.011$). Groundwater total coliform concentrations near the end and front of trenches beneath Field 1 (End: 1071 MPN/100 mL; Front: 518 MPN/100 mL) and Field 2 at WCH (Front: 1140 MPN/100 mL; End: 300 MPN/100 mL) were not significantly different (Figure 5). Overall, pooled data revealed that groundwater near the front of the drainfield trenches had similar geometric mean total coliform concentrations (748 MPN/100 mL) in comparison to groundwater near the end of the trenches (592 MPN/100 mL). Groundwater beneath Field 1 had similar geometric mean coliform concentrations (744 MPN/100 mL) in relation to groundwater beneath Field 2 (585 MPN/100 mL). There was a weak correlation between enterococci and total coliform concentrations at WC ($r = 0.328$, and $p = 0.07$).

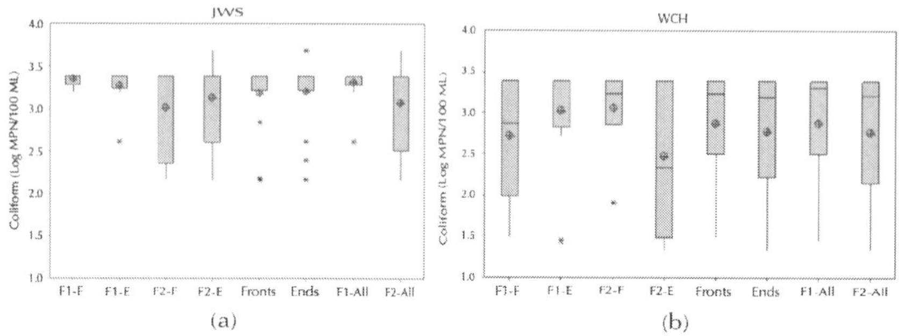

Figure 5: Coliform concentrations (Log MPN/100 mL) in groundwater beneath the drainfields at James W. Smith (a) and West Craven High (b). Field 1 is F1, Field 2 is F2, front of the field is F, end of the field is E. Data indicate that groundwater coliform concentrations were evenly distributed beneath the systems at both sites.

Groundwater E. coli Distribution

The mean groundwater *E. coli* concentrations were higher at JWS near the end of the trenches beneath Field 1 (7 MPN/100 mL) and Field 2 (2 MPN/100 mL) relative to the end of trenches (Field 1 Front: 1 MPN/100 mL, Field 2 Front: 1 MPN/100 mL) (Figure 6). Differences were significant at $p = 0.097$. Groundwater beneath Field 1 had similar average *E. coli* concentrations (4 MPN/100 mL) to groundwater beneath Field 2 (2 MPN/100 mL). The mean groundwater *E. coli* concentrations at WCH towards the front of the trenches for Field 1 (8 MPN/100 mL) and Field 2 (4 MPN/100 mL) were similar to concentrations near the end of the trenches (Field 1 End: 4 MPN/100 mL; Field 2 End: 1 MPN/100 mL) (Figure 6). The pooled data also showed similar mean concentrations of *E. coli* in groundwater towards to the front of the trenches (6 MPN/100 mL) in comparison to the end (3 MPN/100 mL), and beneath Field 1 (6 MPN/100 mL) than Field 2 (3 MPN/100 mL).

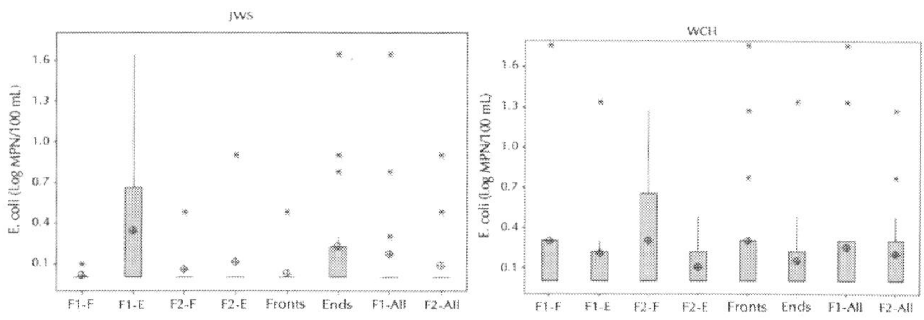

Figure 6: E. coli concentrations (Log MPN/100 mL) in groundwater beneath the drainfield trenches at James W. Smith (a) and West Craven High (b). Field 1 is F1, Field 2 is F2, front of the field is F, end of the field is E. Data indicate that the E. coli concentrations were more variable at James W. Smith relative to West Craven.

Microbial Treatment

The OWS at JWS was efficient (>99%) at reducing all FIB concentrations before discharge to groundwater and adjacent surface water. Wastewater geometric mean FIB concentrations from the septic tank (enterococci: 296,669 MPN/100 mL; total coliform: 1,104,391 MPN/100 mL; and *E. coli*: 457,809 MPN/100 mL) were significantly higher (all $p < 0.05$) than all other sampling locations. Between the JWS septic tank and groundwater beneath the field with the highest FIB concentrations (Field 1), there were significant reductions in the concentrations of enterococci (log 2.65 reduction or 99.78%), total coliform (1og 2.74 reduction or 99.82%) and *E. coli* (log 5.06 reduction or >99.99%) (Figure 7). Groundwater enterococci and total coliform (but not *E. coli*) concentrations beneath portions of Field 1 were significantly elevated relative to background groundwater (enterococci: 105 MPN/100 mL; total coliform: 267 MPN/100 mL; *E. coli*: 0 MPN/100 mL). Microbe concentrations at JWS were further reduced as groundwater moved 24 m down-gradient from the OWS (enterococci: 57 MPN/100 mL; total coliform: 1418 MPN/100 mL; *E. coli:* < 1 MPN/100 mL). Surface water microbe concentrations at the spring down-gradient from the OWS (enterococci: 4 MPN/100 mL; total coliform: 149 MPN/100 mL; *E. coli:* 3 MPN/100 mL) were less than or similar to background groundwater concentrations, and surface water standards. The overall FIB concentration reductions at JWS from septic tank to the spring were significant (enterococci: log 4.86 or 99.98% reduction; total coliform: log 3.87 or 99.87% reduction; and *E. coli*: log 5.18 or >99.99% reduction). Creek water upstream from where the spring discharges into the stream contained elevated concentrations of all microbial indicators (enterococci: 71 MPN/100 mL; total coliform: 1221 MPN/100 mL; and *E. coli*: 291 MPN/100 mL).

The OWS at WCH was also efficient at reducing all measured wastewater FIB concentrations before discharge to groundwater beneath the drainfields (Figure 7). There was a log 2.16 reduction (99.30%) in enterococci concentrations from the septic tank (117,532 MPN/100 mL) to groundwater beneath Field 1 (820 MPN/100 mL). Groundwater enterococci concentrations beneath Field 1 were still significantly (p = 0.03) higher than background groundwater (52 MPN/100 mL). Similar to JWS, the OWS at WCH also had elevated geometric mean enterococci (820 MPN/100 mL) relative to the single sample standard for infrequently used recreational waters (151 MPN/100 mL) [27]. However, there was a log 3.75 reduction (99.98%) between the septic tank and groundwater 10 m down-gradient (21 MPN/100 mL) from the drainfield, and at that distance, enterococci concentrations were not significantly different than background groundwater.

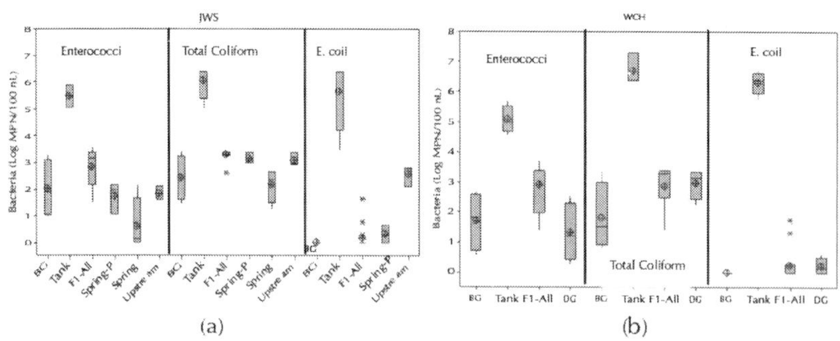

Figure 7: Fecal indicator bacteria concentrations (Log MPN/100 mL) data at James W. Smith (a) and West Craven High (b) for background groundwater (BG), septic tank effluent (Tank), groundwater beneath the drainfield with the highest microbial concentrations (F1-All), down-gradient groundwater (DG), spring water (Spring) and stream (Up-stream). Data indicate excellent treatment (>99% reduction) for all indicator bacteria from the septic tanks

to groundwater, and further treatment as indicator bacteria concentrations decrease in groundwater down-gradient from the systems. Enterococci reductions are the lowest.

There was a log 3.87 reduction (99.98%) in total coliform at WCH between the septic tank (4,859,918 MPN/100 mL) and groundwater beneath Field 1 (744 MPN/100 mL). Total coliform concentrations did not change much 10 m down-gradient from the OWS (949 MPN/100 mL). Background groundwater total coliform concentrations (65 MPN/100 mL) were lower than all other sampling locations, but statistically significant differences were only observed when comparing septic wastewater to background groundwater. Septic wastewater coliform concentrations were significantly ($p < 0.05$) higher than all other sampling locations. The geometric mean total coliform concentrations beneath Field 1 and down-gradient from the OWS were lower than total coliform surface water standards (1000 MPN/100 mL) used by some countries [28].

There was a log 5.63 reduction (>99.99%) in *E. coli* concentrations between the septic tank (2,573,800 MPN/100 mL) and groundwater beneath Field 1 (6 MPN/100 mL), and a log 6.11 reduction (>99.99%) between the tank and groundwater 10 m down-gradient (2 MPN/100 mL) from the OWS at WCH. Septic tank *E. coli* concentrations were significantly higher than all other sampling locations. Groundwater *E. coli* concentrations beneath the OWS were lower than surface water standards for *E. coli* (geometric mean: 126 MPN/100 mL; single sample: 576 MPN/100 mL) [27].

Physical and Chemical Parameters

The mean SC at each site was highest for wastewater samples from the septic tanks (JWS: 1057 ± 387 μS/cm; WCH: 1196 ± 432 μS/cm) (Table 2). Groundwater beneath the drainfields and down-gradient from the OWS had SC significantly ($p < 0.05$) higher than background groundwater for both sites (Table 2). The spring at JWS also had elevated SC (620 ± 69 μS/cm) relative to background groundwater (98 ± 51 μS/cm) and stream water (144 ± 12 μS/cm). Groundwater SC trends were similar to microbial concentration trends at JWS. More specifically, the average groundwater SC was higher beneath Field 1 (755 μS/cm) relative to Field 2 (547 μS/cm), and higher towards the end of the trenches (754 μS/cm) relative to the front (547 μS/cm). At WCH, groundwater SC trends were similar to microbe trends because the average SC was elevated near the front of the trenches (521 μS/cm) relative to the ends of the trenches (340 μS/cm). However, there were some dissimilarities as mean groundwater SC was elevated beneath Field 2 (550 μS/cm) relative to Field 1 (312 μS/cm), while microbe concentrations were typically higher beneath Field 1 at WCH. There were weak but significant correlations between SC and enterococci concentrations at JWS ($r = 0.428$, $p = 0.007$) and WCH ($r = 0.387$, $p = 0.014$) and between SC and total coliform concentrations at WC ($r = 0.534$, $p = 0.001$).

Sample pH values were highest for wastewater at each site (JWS: 7.31 ± 0.26; WCH: 6.91 ± 0.28), and were generally similar to or elevated in groundwater beneath the drainfields (JWS: 6.37 to 6.89; WCH: 5.97 to 6.53) relative to background groundwater (JWS: 5.12 ± 0.65; WCH: 6.56 ± 0.99).

Table 2: Mean and standard deviation (in brackets) physical and chemical properties of water at various sampling locations at James W. Smith and West Craven High. Data indicate higher mean specific conductivity values for groundwater influenced by wastewater

Site	Sampling Location	Specific Conductivity (μS/cm)	pH	Temp (°C)	Depth to Water (m)
JWS	Field 1	753 (246)	6.37 (0.57)	18.3 (1.1)	4.56 (0.91)
	Field 2	547 (237)	6.89 (0.75)	18.9 (1.1)	6.21 (1.02)
	Front	547 (204)	6.45 (0.80)	18.4 (0.9)	4.92 (1.25)
	Ends	754 (268)	6.78 (0.58)	18.8 (1.2)	5.85 (1.15)
	Background	98 (51)	5.12 (0.65)	18.1 (0.5)	4.32 (0.31)
	Tank	1057 (387)	7.31 (0.26)	17.8 (2.9)	-
	Down-gradient	620 (69)	7.23 (0.27)	17.5 (0.8)	0.34 (0.02)
	Spring	445 (29)	6.91 (0.37)	18.2 (0.6)	-
	Upstream	144 (12)	7.22 (0.25)	13.7 (1.1)	-

WCH	Field 1	312 (401)	5.97 (0.65)	16.4 (2.1)	1.76 (0.16)
	Field 2	550 (302)	6.53 (0.34)	16.7 (1.8)	1.62 (0.11)
	Front	521 (327)	6.32 (0.55)	16.6 (2.0)	1.69 (0.16)
	Ends	340 (397)	6.17 (0.63)	16.4 (1.9)	1.69 (0.16)
	Background	49 (12)	6.56 (0.99)	15.5 (2.2)	1.53 (0.31)
	Tank	1196 (432)	6.91 (0.28)	17.9 (3.5)	-
	Down-gradient	710 (212)	6.53 (0.22)	17.2 (1.5)	1.44 (0.13)

Water temperature at JWS was similar for the septic tank wastewater (17.8 ± 2.9 °C), groundwater beneath the drainfields (18.3 to 18.9 °C), and spring (18.2 ± 0.6 °C). The average stream water temperature at JWS was lower (13.7 ± 1.1 °C) than the groundwater and wastewater temperatures.

The average depth to groundwater near the drainfield area at JWS (4.56 to 6.21 m) was greater than at WCH (1.62 to 1.76 m). There was also more variability in groundwater depths between Fields 1 and 2 at JWS (mean difference of 1.55 m) in comparison to WCH (mean difference of 0.14 m).

DISCUSSION

Microbial Distribution

Overall, the OWS at WCH had a more uniform distribution of FIB in groundwater beneath the system than the OWS at JWS. More specifically, statistically significant differences ($p < 0.05$) were observed when comparing the groundwater enterococci concentrations near the end of the trenches relative to the front of the trenches at JWS, but not at WCH. Higher concentrations of enterococci concentrations in groundwater towards the ends of the trenches at JWS may have been because of the formation of a biomat near the front of the trenches that reduced the infiltration rate of effluent and bacteria at JWS near the front [17,29,30,31]. Biomats form along the trench bottom and trench sidewalls as solids, microorganisms, and secretions from microorganisms accumulate and begin to clog soil pores and reduce the infiltration rate of wastewater into the soil. Biomat formation is influenced by, dosing characteristics, hydraulic and organic loading rate, age of the system, and aeration status of the infiltrative surface [30,32,33]. While both OWS used pump stations, and thus had effluent dosing and resting times, the OWS at WCH used low-pressure pipe (LPP) distribution. The LPP system may have provided more even distribution of wastewater across the drainfield area, and limited the development of a biomat relative to JWS, or allowed a more even development of a biomat along the trenches. Prior studies have shown that OWS that use LPP can be effective at reducing microbial concentrations, in theory, because of the uniform distribution of effluent and the resting and dosing cycles which promote unsaturated flow [19,20]. Research has also shown that effluent distribution via distribution box, like at JWS, can often be unequal [5,9]. When effluent leaves the distribution boxes and enters the drainfield trenches at JWS, effluent initially leaks out of the closely spaced, relatively large holes in the pipes at the front of the trenches, thus dosing the front the of the trenches first, and resulting in a relatively high hydraulic and organic loading rate near the front of the trenches [30]. Overtime, a biomat forms at the front of the trenches and gradually extends towards the end, as effluent infiltration is reduced via the biomat and effluent is "pushed" along the length of the trench [30]. The OWS at JWS had been in use for 25 years prior to the start of this study. This may explain why groundwater enterococci, total coliform, *E. coli* concentrations and SC were higher towards the end of the trenches at JWS, because of the biomat formation at the front of the trenches that limited infiltration of wastewater near the front. Biomat formation can enhance the treatment of effluent by reducing the infiltration rate of wastewater into soil, creating unsaturated flow conditions beneath the OWS trenches, and increasing the residence time of wastewater in the vadose zone [30,31,32]. However, if the biomat does not extended to the distal end of the trenches, then uneven distribution and treatment of effluent along the trench may occur, potentially influencing groundwater quality dynamics beneath OWS. Biomat formation takes time. The OWS at WCH and JWS had been is use for 15 years and 25 years respectively, prior to the start of this project, so there was more time for biomat formation at JWS. However, the maximum design wastewater loading rate to drainfield trenches at WCH (66.2 $L/m^2/d$) was greater than at JWS (42.4 $L/m^2/d$), thus increasing the potential for biomat development at WCH. The higher loading rates at WCH were a reflection of the NC regulations for the design and installation of OWS [21]. There were consistent patterns of elevated enterococci, total coliform and *E. coli* concentrations beneath Field 1 relative to Field 2 at both sites. For the JWS site, this is somewhat expected because there was a significant difference in the depth to groundwater beneath Field 1 (4.56 ± 0.91 m) relative to Field 2 (6.21 ± 1.02 m). Therefore, the vadose zone was on average over 1.6 m thicker beneath Field 2 relative to Field 1 at JWS. Prior studies have shown that bacteria treatment typically improves as the unsaturated zone thickness increases [7,8,12,14,15,26]. Therefore, there was likely more bacterial filtration beneath Field 2, relative to Field 1. At WCH, the depth to groundwater was very similar beneath Field 1 (1.76 ± 0.16 m) and Field 2 (1.62 ± 0.11 m). The groundwater samples were collected during a period (November–May) when the water table is typically highest in eastern NC [22]. Therefore, it is possible the treatment efficiencies of the OWS may improve during the summer, when the water table is lower, and vadose zone is thicker.

There were significant differences in groundwater SC near the end of the drainfield trenches at JWS (754 ± 268 µS/cm) relative to the front of the trenches (547 ± 204 µS/cm), and beneath Field 1 (753 ± 246 µS/cm) in comparison to Field 2 (547 ± 237 µS/cm). Because wastewater contains elevated concentration of dissolved salts and solids, the SC of wastewater is elevated relative to most fresh groundwater, and SC has been used as a tracer for wastewater impacted groundwater [34,35]. The SC patterns were very similar to the microbial indicator patterns at JWS (both higher for Field 1 relative to Field 2, and higher towards the end of the trenches relative to front). At WCH, there were also statistically significant differences ($p < 0.05$) in groundwater SC beneath the front (521 ± 327 µS/cm) and ends of the trenches (340 ± 397 µS/cm), and beneath Field 1 (312 ± 401 µS/cm), relative to Field 2 (550 ± 302 µS/cm). However the differences at WCH (Fronts > Ends: $p = 0.050$; and Field 2 > Field 1: $p = 0.011$) were not as significant as at JWS (Ends > Fronts: $p = 0.0167$; Field 1 > Field 2: $p = 0.0070$). Therefore, the groundwater SC also showed a more even distribution at WCH with the LPP system in comparison to JWS with the pump to distribution box system.

While some of the differences in microbial concentrations between Fields 1 and 2 at JWS may have been influenced by the differences in vadose zone thickness, there was still more variation in groundwater microbe concentrations between the front and end of the trenches for each individual field (less variability in vadose zone thickness) at JWS in relation to WCH, indicating that effluent distribution was a key factor in groundwater microbial concentrations.

Treatment

There was significant attenuation (>99%) of all microbial indicators between the septic tank and groundwater for both systems. Total coliform were most abundant in all sampling locations at both sites. The OWS at WCH was on average, more efficient (Log 3.87 or 99.98%) at reducing total coliform concentrations prior to discharge to groundwater than the OWS at JWS (Log 2.74 or 99.82%). However, the OWS at JWS was more efficient at reducing enterococci (Log 2.65 or 99.78%) in comparison to WCH (Log 2.16 or 99.30%). The average *E. coli* treatment at the two sites were similar (JWS: 5.06 or >99.99%; WCH: 5.63 or >99.99%). Overall, the OWS at both sites were less efficient at reducing enterococci, relative to the other bacteria, and thus enterococci was the most resilient of the microbial indicators.

While FIB concentration reduction was efficient, groundwater with elevated concentrations of enterococci and total coliform were still observed beneath the OWS trenches at JWS. For example the geometric mean enterococci concentration beneath F1 (667 MPN/100 mL) was elevated relative to the single sample recreational surface water standard for infrequently used waters (151 MPN/100 mL) [27]. The United States Environmental Protection Agency has not set a surface water standard for total coliform, because coliform bacteria grow in soil and are not specific to fecal material, however, other countries such as Columbia, Cuba, and Japan have total coliform standards for recreational waters of 1000 MPN/100 mL [28]. The geometric mean total coliform concentration beneath JWS (Field 1) was 2007 MPN/100 mL, and thus would have exceeded standards established by other countries. However, FIB concentrations at JWS were further reduced as groundwater moved 24 m down-gradient from the OWS (enterococci: 57 MPN/100 mL; total coliform: 1418 MPN/100 mL; *E. coli:* < 1 MPN/100 mL). Surface water microbe concentrations at the spring down-gradient from the OWS (enterococci: 4 MPN/100 mL; total coliform: 149 MPN/100 mL; *E. coli:* 3 MPN/100 mL) were less than or similar to background groundwater concentrations, and surface water standards. Therefore, FIB transport to surface water was limited at JWS because of the combined effects of the vertical separation from OWS trenches to groundwater and the horizontal setback distances from surface water, resulting in efficient overall treatment (enterococci: Log 4.86 or 99.98% reduction; total coliform: Log 3.87 or 99.87% reduction; and *E. coli*: log 5.18 or >99.99% reduction). Creek water upstream from where the spring discharges into the stream contained elevated concentrations of all FIB (enterococci: 71 MPN/100 mL; total coliform: 1221 MPN/100 mL; and *E. coli*: 291 MPN/100 mL) indicating another source of surface water FIB contamination in the watershed.

Groundwater typically had higher enterococci concentrations than *E. coli*, but surface waters typically had higher *E. coli* relative to enterococci. Prior studies have also reported that OWS typically are not as effective at reducing enterococci relative to *E. coli* [8,9,10]. Transport of FIB beyond 10 m from the OWS was very limited, and groundwater FIB concentrations were typically at or below background concentrations 10 to 24 m down-gradient from the OWS. While the spring at JWS did contain detectable levels of all FIB (Enterococci: 4 MPN/100 mL; Total Coliform: 222 MPN/100 mL; *E. coli*: 2 MPN/100 mL), the mean spring FIB concentrations were lower than surface water sampled 5 m upstream from where the spring discharged into the creek (Enterococci: 71 MPN/100 mL; Total Coliform: 1221 MPN/100 mL; *E. coli*: 291 MPN/100 mL). Therefore, other sources of FIB pollution may have been influencing stream water quality. Research has shown that pets [36], livestock [37], wildlife [38], and various other non-point sources of pollution [39] can contribute significant concentrations of FIB to surface waters. There are some livestock farms upstream from JWS, sparse residential development with OWS, and wildlife such as raccoons, birds, and deer have been seen in riparian areas adjacent to the stream. Studies have also shown that surface waters near OWS may have FIB concentrations that are elevated relative to groundwater near properly functioning and malfunctioning systems [8,9,10].

CONCLUSIONS

Overall, there was a more even distribution of FIB in groundwater beneath the LPP system at WCH in comparison to the pump to distribution box system at JWS. The groundwater SC was also more similar beneath the drainfield at WCH in relation to JWS. Both systems were very effective (>99%) at reducing FIB concentrations before discharge to groundwater, and FIB concentrations were typically at background groundwater concentrations 10 to 24 m down-gradient from the OWS. The OWS at both sites were less efficient at reducing enterococci concentrations in relation to *E. coli* and total coliform treatment.

ACKNOWLEDGMENTS

The authors would like to thank the North Carolina Water Resources Research Institute and the East Carolina University Coastal Water Resources Center for partial funding of this research. The authors would also like to acknowledge the assistance that John Woods, Jim Watson, Sarah Hardison, Matt Smith, Eliot Anderson-Evans, Caitlin van Dodewaard, and Amberlynne VanDusen provided with field and/or lab work and data compilation.

AUTHOR CONTRIBUTIONS

All authors contributed to the conception, field and/or lab work, and development of this manuscript.

CONFLICTS OF INTEREST

The authors declare no conflict of interest.

REFERENCES

1. United States Environmental Protection Agency (USEPA). *Onsite Wastewater Treatment Systems Manual*; EPA/625/R-00/008. USEPA: Washington, DC, USA, 2002; pp. 1–11.
2. Carroll, S.; Hargreaves, M.; Goonetilleke, A. Sourcing faecal pollution from onsite wastewater treatment systems in

surface waters using antibiotic resistance analysis. *J. Appl. Microbiol.* 2005, *99*, 471–482.

3. Harman, J.; Robertson, W.D.; Cherry, J.A.; Zanini, L. Impacts on a sand aquifer from an old septic system: Nitrate and phosphate. *Groundwater* 1996, *34*, 1105–1114.
4. Pang, L.; Close, M.; Goltz, M.; Sinton, L.; Davies, H.; Hall, C.; Stanton, G. Estimation of septic tank setback distances based on transport of *E. coli* and F-RNA phages. *Environ. Int.* 2003, *29*, 907–921.
5. Patel, T.; O'Luanaigh, N.O.; Gill, L.W. A Comparison of gravity distribution devices used in On-site domestic wastewater treatment systems. *Water Air Soil Pollut.* 2008, *191*, 55–69.
6. Lowe, K.S.; Rothe, N.K.; Tomaras, J.M.B.; DeJong, K.; Tucholke, M.B.; Drewes, J.; McCray, J.E.; Munakata-Marr, J. Influent constituent characteristics of the modern waste stream rom single sources. In *Literature Review*; Water Environment Research Foundation: Alexandria, VA, USA, 2007; pp. 3–19.
7. Humphrey, C.P.; O'Driscoll, M.A.; Zarate, M.A. Evaluation of on-site wastewater system Escherichia coli contributions to shallow groundwater in coastal North Carolina. *Water Sci. Technol.* 2011, *63*, 789–795.
8. Conn, K.E.; Habteselassie, M.Y.; Blackwood, A.D.; Noble, R.T. Microbial water quality before and after the repair of a failing onsite wastewater treatment system adjacent to coastal waters. *J. Appl. Microbiol.* 2011, *112*, 214–224.
9. Habteselassie, M.Y.; Kirs, M.; Conn, K.E.; Blackwood, A.D.; Kelly, G.; Noble, R.T. Tracking microbial transport through four onsite wastewater treatment systems to receiving waters in eastern North Carolina. *J. Appl. Microbiol.* 2011, *111*, 835–847.
10. Harris, J.; Humphrey, C.; O'Driscoll, M. Transport of indicator microorganisms from an onsite wastewater system to adjacent stream. *Univers. J. Environ. Res. Technol.* 2013, *3*, 423–426.
11. Cahoon, L.B.; Hales, J.C.; Carey, E.S.; Loucaides, S.; Rowland, K.R.; Nearhoof, J.E. Shellfishing closures in southwest Brunswick County, North Carolina: Septic tanks *vs.* stormwater runoff as fecal coliform sources. *J. Coast. Res.* 2006, *22*, 319–327.

12. Meeroff, D.; Bloetscher, F.; Bocca, T.; Morin, F. Evaluation of water quality impacts of on-site wastewater treatment and disposal systems on urban coastal waters. *Water Air Soil Pollut.* 2008, *192*, 11–24.

13. Borchardt, M.A.; Po-Huang, C.; DeVries, E.O.; Belongia, E.A. Septic system density and infectious diarrhea in a defined population of children. *Environ. Health Perspect.* 2003, *111*, 742–748.

14. Scandura, J.E.; Sobsey, M.D. Viral and bacterial contamination of groundwater from on-site sewage treatment systems. *Water Sci. Technol.* 1997, *35*, 141–146.

15. Karathanasis, A.D.; Mueller, T.G.; Boone, B.; Thompson, Y.L. Effect of soil depth and texture on fecal bacteria removal from septic effluents. *J. Water Health* 2006, *4*, 395–404.

16. Ijzerman, M.M.; Hagedorn, C.; Reneau, R.B. Microbial tracers to evaluate an on-site shallow-placed low pressure distribution system. *Water Res.* 1993, *27*, 343–347.

17. O'Luanaigh, N.D.; Gill, L.W.; Misstear, B.D.R.; Johnston, P.M. The attenuation of microorganisms in on-site wastewater effluent discharged into highly permeable subsoils. *J. Contam. Hydrol.* 2012, *142–143*, 126–139.

18. Motz, E.C.; Cey, E.; Ryan, M.C.; Chu, A. Vadose zone microbial transport below at-grade distribution of wastewater effluent. *Water Air Soil Pollut.* 2012, *223*, 771–785.

19. Carlile, B.L.; Cogger, C.G.; Sobsey, M.D.; Scandura, J.; Steinbeck, S.J. *Movement and Fate of Septic Tank Effluent in Soils of the North Carolina Coastal Plain*; Report to the Coastal Plains Regional Commission: Raleigh, NC, USA, 1981; p. 37.

20. Ijzerman, M.M.; Hagedorn, C.; Reneau, R.B. Fecal indicator organisms below an on-site wastewater system with low pressure pipe distribution. *Water Air Soil Pollut.* 1992, *63*, 201–210.

21. North Carolina Division of Environmental Health: On-site Wastewater Section. *Laws and Rules for Sewage Treatment and Disposals Systems*; Raleigh, NC, USA, 1999. Available online: http://ehs.ncpublichealth.com/oswp/resources.htm (accessed on 11 March 2014).

22. U.S. Department of Agriculture, Natural Resources Conservation

Service. Soil Survey of Craven County, North Carolina, USA. 1989. Available online: http://soils.usda.gov/survey/printed_surveys/ (accessed on 11 March 2014).

23. Blott, S.; Pye, K. Gradistat: A grain size distribution and statistics package for the analysis of uncosolidated sediments. *Earth Surf. Process. Landf.* 2001, *26*, 1237–1248.
24. Domenico, P.A.; Schwartz, W. *Physical and Chemical Hydrogeology*, 2nd ed.; John Wiley & Sons, Inc.: New York, NY, USA, 1998; p. 36.
25. Del Rosario, K.L.; Humphrey, C.P.; Mitra, S.; O'Driscoll, M.A. Nitrogen and carbon dynamics beneath on-site wastewater treatment systems in pitt county, North Carolina. *Water Sci. Technol.* 2014, *69*, 663–671.
26. Humphrey, C.P.; O'Driscoll, M.A. Biogeochemistry of groundwater beneath on-site wastewater systems in a coastal watershed. *Univers. J. Environ. Res. Technol.* 2011, *1*, 320–328.
27. United States Environmental Protection Agency (USEPA). *Managing Urban Watershed Pathogen Contamination*; EPA/600/R-03/111. USEPA: Cincinnati, OH, USA, 2003.
28. United States Environmental Protection Agency. *Ambient Water Quality Criteria for Bacteria*; EPA440/5–84–002. USEPA: Washington, DC, USA, 1986.
29. Stevik, T.K.; Aa, K.; Ausland, G.; Hanssen, J.F. Retention and removal of pathogenic bacteria in wastewater percolating through porous media: A review. *Water Res.* 2004, *38*, 1355–1367.
30. Hoover, M.T.; Disy, T.A.; Pfieffer, M.A.; Dudley, N.; Mayer, R.B.; Buffington, B. *North Carolina Subsurface Wastewater Operators Training School Manual*; 1996.
31. Beal, C.D.; Gardner, E.A.; Kirchhof, G.; Menzies, N.W. Long-term flow rates and biomat zone hydrology in soil columns receiving septic tank effluent. *Water Res.* 2006, *40*, 2327–2338.
32. Siegrist, R.; Boyle, W.C. Wastewater-induced soil clogging development. *J. Environ. Eng.* 1987, *113*, 550–566.
33. Stevik, T.K.; Ausland, G.; Jenssen, P.D.; Siegrist, R. Removal of *E. coli* during intermittent filtration of wastewater effluent as affected by dosing rate and media type. *Water Res.* 1999, *33*, 2088–2098.
34. Alhajjar, B.J.; Chesters, G.; Harkin, J.M. Indicators of Chemical

Pollution from Septic Systems. *Groundwater* 1990, *28*, 559–568.

35. Humphrey, C.P.; Deal, N.E.; O'Driscoll, M.A.; Lindbo, D.L. Characterization of on-site wastewater nitrogen plumes in shallow coastal aquifers, North Carolina. In Proceedings of the 2010 World Environmental & Water Resources Congress, Providence, RI, USA, 16–20 May 2010; pp. 949–958.
36. Wright, M.E.; Solo-Gabriele, H.M.; Elmir, S.; Fleming, L.E. Microbial load from animal feces at a recreational beach. *Mar. Pollut. Bull.* 2009, *58*, 1649–1656.
37. Liwimbi, L.; Graves, A.K.; Israel, D.W.; van Heugten, E.; Robinson, B.; Cahoon, C.W.; Lubbers, J.F. Microbial source tracking in a watershed dominated by swine. *Water* 2010, *2*, 587–604.
38. Whitlock, J.E.; Jones, D.T.; Harwood, V.J. Identification of the sources of fecal coliforms in an urban watershed using antibiotic resistance analysis. *Water Res.* 2002, *36*, 4237–4282.
39. Sanders, E.C.; Yuan, Y.; Pitchford, A. Fecal coliform and *E. coli* concentrations in effluent-dominated streams of the Upper Santa Cruz Watershed. *Water* 2013, *5*, 243–261.

Chapter 7

Performance of an Anaerobic Baffled Filter Reactor in the Treatment of Algae-Laden Water and the Contribution of Granular Sludge

Yaqin Yu[1,2], Xiwu Lu[1] and Yifeng Wu[1]

[1]School of Energy and Environment, Southeast University, Nanjing 210096, China;
[2]Department of Civil Engineering, Yancheng Institute of Technology, Yancheng 224003, China

ABSTRACT

This study investigated the performance and stability of an anaerobic baffled filter reactor in the treatment of algae-laden water from

Taihu Lake at several organic loading rates. The study also evaluated the capability of soft filler to train granule sludge and improve the anaerobic environment and sludge activity in the anaerobic baffled reactor (ABR), thereby enhancing the treatment efficiency. The ABR consisted of five rectangular compartments, each of which was 120 cm long, 80 cm wide, 80 cm high, and packed with soft filler. The anaerobic baffled filter reactor was found to be an efficient reactor configuration for the treatment of algae-laden water. The reactor was operated at an organic loading rate of 1.5 kg chemical oxygen demand (COD)/(m^3d) and an ambient temperature of 30 °C; under these conditions, the COD removal efficiency was 80% and the biogas production rate was 293 mL/(Ld). Moreover, the soft filler increased the biomass retention time and decreased the rate at which solids were washed out from the reactor, promoting an improved spatial distribution of the microbial communities within the compartments. *Methanoregula*, *Methanobacteriaceae*, *Methanosaeta*, *Methanoculleu*, and *Thermogymnomonas* were the dominant archaeal species in each compartment during an operational period of approximately 100 days. The protease activity in the reactor decreased longitudinally down the reactor from Compartments 1 to 5, whereas the activity of coenzyme F_{420} increased. The soft filler played a key role in successfully treating algae-laden water with the anaerobic baffled filter reactor.

INTRODUCTION

The presence of excess nutrients, especially phosphorus and nitrogen, in lakes and reservoirs has been a major factor stimulating blue-green algal blooms [1]. Taihu Lake, which is the third largest lake in China and a source of drinking water for more than two million people in Jiangsu Province, has exhibited serious water toxicity problems due to eutrophication [2]. Water pollution caused by the excessive growth of blue-green algae has become a growing environmental problem. One current approach to reducing the algae in Taihu Lake is to refloat the algae after a bloom has occurred [3,4]. A salvage ship takes up water-laden algae in quantities that can reach several thousand cubic meters per day. However, these large quantities of algae-laden water could have serious secondary environmental effects, including the pollution of groundwater, air, and soil without sufficing and effective algae-laden water treatments [5].

Blue-green algae cells are typically rich with polysaccharides and contain a lower cellulose concentration than lignin. Furthermore, blue-green algae cells can be easily degraded to methane during anaerobic digestion processes. Therefore, anaerobic digestion offers key advantages in the treatment of algae-laden water, such as energy production and pollutant removal. Furthermore, the biogas generation rate of algae degradation given in previous research is typically more feasible than that of other recovery methods, such as incineration and composting [5,6].

The design and operation of reactors has been studied to develop anaerobic biotechnology for algae-laden water treatment. Various anaerobic reactors, such as the anaerobic filter (AF), up flow anaerobic sludge blanked reactor (UASB), anaerobic continuous stirred tank reactor (CSTR), and anaerobic baffled reactor (ABR), have been tested and shown to be effective in the literature. In particular, the ABR offers several advantages over traditional reactors, including granulation promotion and phase separation, because of the reactor's unique design characteristics. The ABR also allows acidified hydrolyzation to predominate in the anterior compartment section and methanogenesis to dominate in the subsequent section [7,8].

Scum consisting of suspensions algae is easily formed in digesters, which increases the risk of algae agglomeration and clogging due to the gas-filled sacs in algae. Filling an ABR with a soft filler, such as assembled fiber filler, can prevent the floatation of algae in the reactor.

Previous studies have demonstrated that source water can be purified in a eutrophic lake using indigenous enrichment microbes with a soft filler. Such a purification system was used in Taihu Lake, and the enrichment microbes on the soft filler were found to effectively degrade the lake's algae and microcystins [9]. Biomass in the reactor may be further retained with the addition of soft filler, which could promote biofilm formation and granular sludge cultivation. Therefore, this study aimed to explore the feasibility and performance of an ABR in treating algae-laden water and to determine the capability of soft filler to enhance granule formation and promote phase separation in the ABR during the treatment of algae-laden water.

MATERIALS AND METHODS

Experimental Devices

A 80-cm-wide, 120-cm-long, 80-cm-high ABR was constructed using polyvinylchloride to treat algae-laden water at an active volume of 740L. The ABR consisted of five rectangular compartments, labeled Compartments 1, 2, 3, 4 and 5; the volumes of these compartments were 200, 180, 120, 120, and 120 L, respectively. Downflow and upflow chambers were created within each compartment using an additional baffle. The widths of the downflow and upflow chambers of Compartments 1 and 2 were 10 and 45 cm, respectively. The widths of the downflow and upflow chambers of Compartments 3, 4, and 5 were 9 and 24 cm, respectively. The five compartments were packed with soft filler made of fiber filler with a diameter of 2 cm. The soft filler was made by the Yixing Dushan Environmental Protection Equipment Company. This soft filler had a high specific surface area of up to 2000 m^2/m^3. Furthermore, the filler enabled biomass retention by attachment and was able to entrap algae from the algae-laden water. The lower parts of the hanging baffles were angled at 45° to evenly direct the flow into the center of the upflow chamber. This configuration facilitated effective mixing and contact between the wastewater and anaerobic sludge at the base of each upflow chamber [10]. Each compartment was equipped with sampling ports that allowed biological solids, gas, and liquid samples to be withdrawn. The liquid sampling ports were located 100 mm from the top of each compartment and the sludge sampling ports were at the bottom of each compartment. Biogas was collected separately via a porthole in the top of the reactor. During the experimental period, the volumes of biogas in the compartments were measured daily using a wet gas meter (Model LML-1, Changchun Filter Co., Ltd., Changchun, China), and the influent feed to the ABR was controlled by a peristaltic pump (Model BT100-01, Baoding, China). The reactor was housed in a temperature-controlled room maintained at 30 °C. Figure 1 provides a schematic of the anaerobic baffled filter reactor.

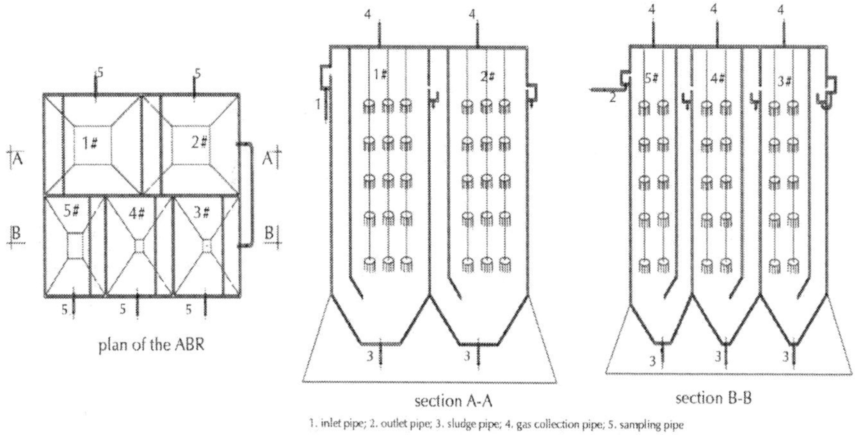

Figure 1: Schematic of the anaerobic baffled filter reactor.

Wastewater Characteristics

The algae-laden water used in this investigation was freshly collected from a blue-algae refloating site near Taihu Lake, Yixing City, Jiangsu Province, China, at the end of June 2012. The raw water, with a pH of 5.8–7.3, had a chemical oxygen demand (COD) of 3000–7000 mg/L, total suspended solids (TSS) of 365–1002 mg/L, total nitrogen (TN) of 55–125 mg/L, and total phosphorus (TP) of 125–183 mg/L. The algae-laden water was diluted using water from Taihu Lake to reach the required concentration.

Seed Sludge

The ABR was initially seeded with sewage sludge, which was collected from a UASB used to treat algae-laden water, thus shortening the cultivation time. Debris and large particles were removed from these wage sludge with a 3-mm sieve. The ratio of mixed liquor volatile suspended solids (MLVSS) to mixed liquor suspended solids (MLSS) in the inoculated sewage sludge was 0.75; the sewage sludge was introduced uniformly into the ABR's five compartments. The sewage sludge concentration in each compartment was 28.5 g MLVSS/L after inoculation.

Operation of the Anaerobic Baffled Reactor

After inoculation, the reactor was sealed and the headspace above each compartment was flushed with nitrogen gas to displace residual air from the system. Initially, the algae-laden water was diluted with water from Taihu Lake to reduce the COD. The reactor was allowed to stabilize for 10 days before the experiments were started to stimulate biofilm growth on the soft filler in each compartment. The experiment performed in this study was then divided into three stages. During each stage, the steady state was marked by relatively stable effluent pH values and COD values, with variations of less than 5%. At the beginning of the start-up, the reactor was operated at low organic loading to acclimate the sludge. The hydraulic retention time (HRT) was held constant at 10 days, and the concentration of the feed was varied from 500 to 2500 mg/L. The reactor was then operated for 60 days during the second stage to investigate the performance. In this stage, an organic loading rate (OLR) ranging from 0.26 to 1.2 kg COD/(m^3d) was achieved by increasing the influent concentration to 6000 mg/L and decreasing the HRT from seven to five days. A shock experiment was completed within one month during the third stage. The feed's COD concentration was increased to 7200 mg/L to determine the stability of the reactor treating the algae-laden water.

Analytical Methods

Chemical Analysis

The total COD, MLSS, MLVSS, pH, and oxidation-reduction potential (ORP) were measured using standard methods [11]. The total volatile fatty acids (VFAs) were detected in membrane-filtered samples (0.45 μm) using gas chromatography (GC, Agilent 4890D) with a flame ionization detector (FID) and a 2-m stainless column packed with Porapak GDX103 (60/80 mesh) [11]. Sludge samples were collected from each compartment of the reactor after stable operation for 90days, and the biomass was examined using scanning electron microscopy (SEM) [12].

The biogas composition (CH_4, CO_2, and N_2) was analyzed using gas chromatography with a multigas analyzer (Model GC-2001) [13]. The COD, pH, ORP, and biogas yield were monitored daily, and the enzymatic activity in the sludge was measured once per week during the experimental period. The VFAs and biogas constituents were measured when the reactor reached steady-state conditions.

Determination of Protease and Co-Enzyme F_{420} Activity

The protease and coenzyme F_{420} activities in the sewage sludge were measured by ultraviolet (UV) spectrophotometry. The protease activity of the sewage sludge was measured according to the procedures given by Whiteley *et al.*, [14]. The sludge (1.0 mL) was incubated at 37 °C for 1 h in a buffer (0.02 mol/L Na_2HPO_4, 0.01 mol/L $C_6H_8O_7$) at pH 6 with 2% azocasein (1.0 mL). Ice-cold 10% (w/v) trichloroacetic acid (TCA) (1.0 mL) was added to stop the reaction, and the tubes were left at −20 °C to facilitate precipitation. The precipitated protein was removed by centrifugation (4000 × g for 10 min), and the A_{440} nm of the TCA-soluble peptides was measured. The coenzyme F_{420} activity of the sewage sludge was measured according to the procedures given by Whitmore *et al.*, [15]. The sludge was suspended in 0.5 M NaCl after being centrifuged, and the pellet was suspended in water (10 mL) and the cells disrupted by heating in a water bath at 100 °C for 10 min. After cooling, the suspension was centrifuged (17,000 × g for 10 min), and the supernatant fluid was diluted 1:2 with 0.3 M NaC1 in 50 n-Tris-HC1 buffer (pH 7.5). The solution was then applied to a QAE-Sephadex A-25 column. The eluate was collected, and the absorbance spectrum was measured at 420 nm with a dual beam spectrophotometer.

DNA Analysis

The predominant methanogens in the sludge samples were determined using denaturing gradient gel electrophoresis (DGGE) [16]. The total DNA from the mixed culture was extracted using the phenol-chloroform method and estimated by 1% agarose gel electrophoresis [17]. The total extracted genomic DNA was used as a template to amplify the V4-5 variable region on 16S rRNA by PCR. The nucleotide sequences of the universal primers were as follows: primer F787, 5′- CGC CCG CCG CGC GCG GCGGGCGGG GCG

GGGGCACGGGGG G CGATTAGATACCCSBGTAGTCC -3′; R1059 (5′- GCCATGCACCWCCTCT -3′) [18]. The PCR amplification was performed in an automated thermal cycler (Bio-Rad, Hercules, California, USA). Initial denaturation was performed at 94 °C for 4 min, followed by 30 cycles of denaturation at 94 °C for 30 s and annealing at 56 °C for 1 min. Each annealing step was followed by extension at 72 °C for 1 min, except for the final 7-min extension step. The PCR products were analyzed by electrophoresis on a 1.5% (w/v) agarose gel for 30 min at 120 V with ethidium bromide in 1× TAE buffer (40 mMTris acetate, 1 mM EDTA, pH 8.0) [16]. Images of the gel were recorded to determine the size, purity, and concentration of the DNA used by the gene genius bioimaging system [16,17].

DGGE analysis was performed using the DCode Universal Mutation Detection System (Bio-Rad, Hercules, California, USA). The PCR samples were analyzed directly on 8.0% (w/v) polyacrylamide gels in 1 × TAE buffer (40 mMTris acetate, 1 mM EDTA, pH 8.0) with a denaturing gradient that ranged from 30% to 60% (100% denaturant is 7 M urea and 40% v/v formamide). The gradient gel was cast with a gradient delivery system. Electrophoresis was performed at a constant voltage of 180 V at 60 °C. After 4 h of electrophoresis, the gel was stained with ethidium bromide for 30 min and visualized on a UVI transilluminator (U-3010 Hitachi, Tokyo, Japan).

The bands in the gel were excised and rinsed in 1 mL of deionized water. The gel was crushed in 20 μL of sterile deionized water and incubated at 30 °C for 1 h. The DNA extract was used as a template in the PCR amplification, as described above, and the PCR products were sequenced on an ABI LR-377 DNA sequencer [16]. The sequences of successfully amplified DNA fragments were then compared with the Gen Bank database using the basic local alignment search tool (BLAST) analysis [16].

RESULTS AND DISCUSSION

Performance of the ABR

COD Removal

The performance of the anaerobic baffled filter reactor was evaluated by the COD removal efficiency throughout the operation of the reactor. The substrate degradation pattern varied as a function of the OLR over an operational period of more than 130 days. Figure 2 presents the COD concentrations of the influent and effluent of each compartment and the COD removal efficiencies during various operations over the study period. During the initial start-up stage, the reactor was operated at a lower OLR to train the sludge to improve the anaerobic environment and sludge activity. The COD removal efficiency was approximately 75% for sludge acclimatization. This result may have been obtained because the organic matter of the influent was absorbed by the biomass of the soft filler in the initial stage. Up to 30 days of operation, when the influent with a COD concentration of 2000–2500 mg/L was applied to the ABR, the effluent COD concentration of the final compartment was lower than 550 mg/Land and the COD removal efficiency ranged from 75% to 80%. A gradual improvement in algal degradation was observed, indicating biofilm and granular sludge formation in the reactor. The reactor was operated in the steady state after the anaerobic activated sludge acclimatized over 40 days. Approximately 80% COD removal occurred when the organic loading rate ranged from 0.26 to 1.2 kg COD/(m^3d). Most of the influent COD was removed in Compartments 1 (34%) and 2 (40%); a smaller percentage of the influent COD was removed in Compartments 3 (25%) and 4 (23%), and Compartment 5 removed the remaining fraction of the influent COD (13%). During the shock stage, temporary acidification occurred in Compartments 1 and 2 during the initial period (Days 102–110) due to the increasing organic loading rate; however, the COD removal of the reactor did not decrease. The COD concentration of the effluent was less than 1500 mg/L after seven days, indicating that the anaerobic baffled filter reactor possessed greater ability to treat the algae-laden water than indicated by the research by Jing [19]. Jing *et al.* have disposed of algae with an ABR and obtained an effluent COD concentration of 3500 mg/L and a removal rate of 55%. In this study, absorption played a dominant role in improving the effluent quality. At first, the organic matter was absorbed on the surface of the soft filler. When the biomass on the soft filler's packing was growing well with high activity, organic matter was absorbed on the algal biomass, which became attached to the soft filler.

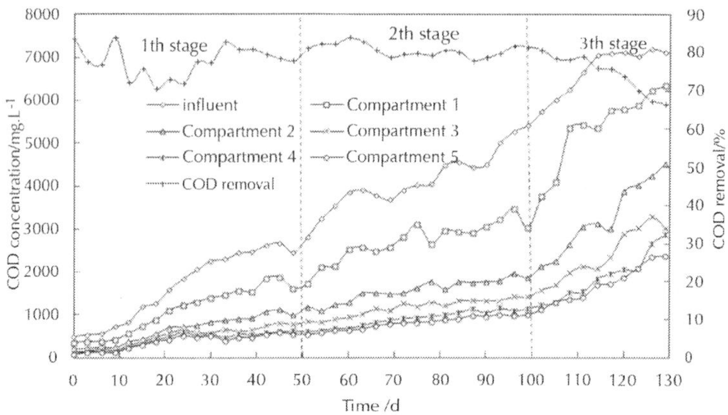

Figure 2: Chemical oxygen demand (COD) removal contributed by each compartment of the reactor.

Post-treatment, such as biological contact oxidation and aquatic vegetation wetland use, was studied based on the results of treating the aerobically digested effluent from algae in Taihu Lake. The results indicated that the contact oxidation tank reduced the inlet load of the subsequent aquatic vegetation wetland, and the COD of the average effluent of the combined process was less than 50 mg/L, which satisfied Chinese standard GB18918-2002 for the discharge of pollutants [20].

In the ABR filled with soft filler, Compartments 1 and 2 were important to the treatment of the algae-laden water. These two compartments acted as a buffer zone to all toxic and inhibitory material in the feed, allowing the subsequent compartments to be loaded with a relatively harmless, balanced, and primarily acidified influent [21]. Occasionally, the soft filler played a critical role in intercepting the algae, which prevented both the algae from floating off and encrustation in the compartment.

Biogas Production

The performance of the anaerobic reactor was measured in terms of the observed biogas yield and methane content. As shown in Figure 3, the daily total gas production rates were low, with an OLR of 0.05 kg COD/(m^3d) during the start-up phase. When the system formed granular sludge during long domestication and cultivation, the biogas production increased with increasing OLR. The biogas production rates varied from 99 to 293 mL/(Ld), with OLRs ranging from 0.26 to 1.50 kg COD/(m^3 d).

Gas samples were collected and analyzed every five days during the steady-state phase. In this phase, the gas composition was generally stable, with CO_2 concentrations in Compartments 1 and 2 of approximately 45%–50% and CH_4 concentrations of 40%–48%. In Compartments 3, 4 and 5, the CO_2 concentration decreased to 30% and the CH_4 concentration increased to 55%–64%. This result demonstrated the ABR's remarkable phase separation characteristics. In the former compartments, the hydrolysis-acidification reactions of the algae predominated, whereas methanogenesis occurred in the latter compartments. These results demonstrate the feasibility of using algae to produce biogas in an anaerobic baffled filter reactor with microorganisms.

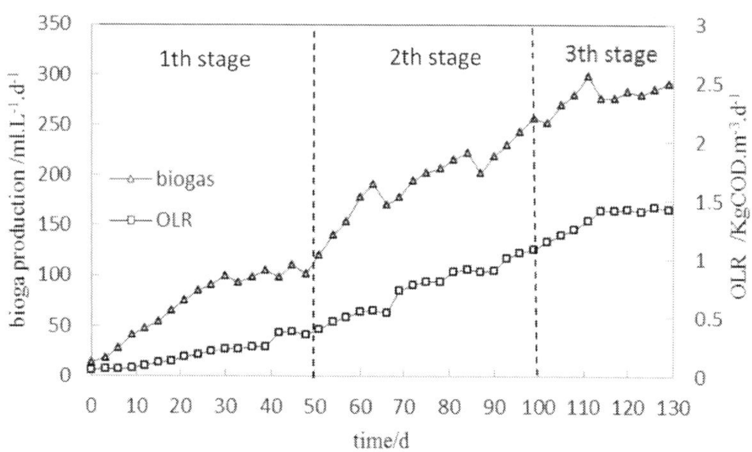

Figure 3: Impact of the organic loading rate (OLR) on biogas production.

Microcystin Removal

Microcystins exist as intracellular compounds in healthy cyanobacteria but are released upon cell lysis [22]. Microcystins pose a health risk to animal and human populations when surface water is used for drinking during post-bloom periods [23,24]. The total and extracellular microcystins were monitored each week to demonstrate the efficacy of the anaerobic baffled filter reactor in degrading microcystins.

During the algal bloom, the concentration of microcystins in the algae-laden water salvaged from Taihu Lake was elevated, with concentrations of up to 500 µg/L. The concentrations of total microcystin-LR (TMC-LR) and extracellular microcystin-LR (EMC-LR) in the algae-laden water ranged from 326 µg/L to 532 µg/L and 204 µg/L to 321 µg/L, respectively. This result indicated that Taihu Lake had been heavily polluted by microcystins due to eutrophication. Figure 4 and Figure 5 illustrate the ability of the reactor to remove TMC-LR and EMC-LR, respectively. The removal rate of TMC-LR in the ABR filled with soft filler ranged from 87.6% to 94.6%, whereas the removal rate of EMC-LR ranged from 77.4% to 91.7%. The ABR removed TMC-LR more extensively than EMC-LR, which indicated that dissolved TMC-LR had a superior biodegradation ability compared to EMC-LR.

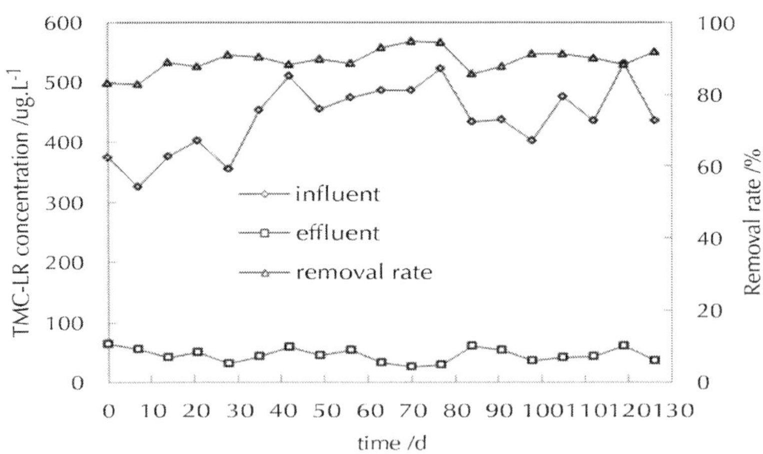

Figure 4: Total microcystin-loading rate (TMC-LR) removal by the anaerobic baffled filter reactor.

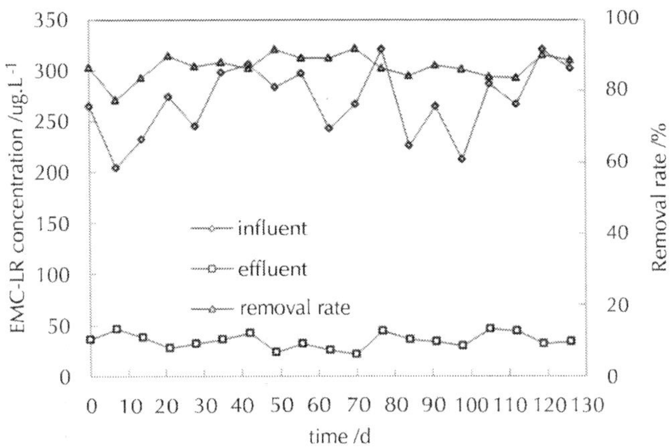

Figure 5: Extracellular microcystin-loading rate (EMC-LR) removal by the anaerobic baffled filter reactor.

The results of this experiment indicated that the anaerobic baffled filter reactor efficiently removed both TMC-LR and EMC-LR. Because the soft filler had a high specific surface area, it could adsorb and retain more algae in the algae-laden water. In other words, the soft filler increased the HRT of the algae and the total quantity of algae in the filler, allowing additional algae to be removed in the reactor.

PH, VFAs, and ORP

PH, VFAs, and ORP are the key parameters affecting both the anaerobic treatment efficiency and microbial community during anaerobic digestion [25,26]. Table 1 provides the pH and ORP levels, and VFA concentrations in each compartment under various operating conditions. The pH of the influent fluctuated from 6.35 to 7.05 during the reaction period. In Compartment 1, the pH dropped from 6.35 to 5.70 over the three stages. However, the pH levels of Compartments 2, 3, 4 and 5 shifted from 6.76, 7.13, 7.15, and 7.20 to 6.60, 7.10, 7.25, and 7.40, respectively. In this study, the relatively low pH values in Compartments 1 and 2 indicated that the substrates in the anaerobic digestion had hydrolyzed to simple organics, which were fermented to VFAs by the acidogens, acidogenesis, and acetogenesis occurring in these compartments. The pH values in Compartments 3, 4 and 5 were maintained at constant levels due to the effective consumption of VFAs. This result demonstrated that the VFAs had been converted to acetate and H_2 gas by acetogens, and these intermediates were then converted to CH_4 by methanogens in the subsequent compartments [27,28]. The anaerobic baffled filter reactor configuration promoted phase separation, and reactor souring did not occur at any time during the operating period.

The ORP, which is a factor in anaerobic digestion, affects the activity of methanobacteria. As shown in Table 1, the ORP value decreased in every compartment over the three stages, from approximately −190 mV in Compartment 1 to approximately −312 mV in Compartment 5 during the first stage. During the third stage, the ORP decreased from −252 mV in Compartment 1 to −341 mV in Compartment 5. The optimal ORP for methane oxidizing activity is approximately −330 mV [26]. The ORP values in these compartments had previously been thought to be maintained at a constant level without manual intervention.

The highest VFA concentration was observed in the first compartment and increased in every compartment. The VFA concentration also decreased longitudinally along the reactor. This behavior demonstrated that hydrolysis and acidogenesis were the main biochemical activities occurring in the first few compartments, whereas methanogenesis dominated in the last few compartments.

Table 1: Variation of pH, volatile fatty acids (VFAs), and oxidation-reduction potential (ORP) in the (anaerobic baffled reactor) ABR.

Stages	Indexes	Influent	Compartment 1	Compartment 2	Compartment 3	Compartment 4	Compartment 5
First stage steady state	pH	6.70 ± 0.35	6.35 ± 0.23	6.76 ± 0.25	7.13 ± 0.34	7.15 ± 0.25	7.20 ± 0.23
	VFAs (mg/L)	—	800 ± 65	650 ± 43	325 ± 35	227 ± 25	205 ± 27
	ORP (mV)	—	−190 ± 34	−250 ± 21	−284 ± 15	−302 ± 17	−312 ± 13
Second stage steady state	pH	6.70 ± 0.35	6.05 ± 0.21	6.80 ± 0.17	7.15 ± 0.07	7.28 ± 0.06	7.30 ± 0.05
	VFA (mg/L)	—	1800 ± 165	970 ± 84	440 ± 38	320 ± 34	170 ± 25
	ORP (mV)	—	−247 ± 25	−297 ± 18	−313 ± 21	−318 ± 15	−325 ± 11
Third stage steady state	pH	6.70 ± 0.35	5.70 ± 0.19	6.60 ± 0.16	7.10 ± 0.08	7.25 ± 0.03	7.40 ± 0.03
	VFA (mg/L)	—	4450 ± 187	1740 ± 93	680 ± 72	480 ± 32	240 ± 26
	ORP (mV)	—	−252 ± 23	−317 ± 19	−325 ± 12	−334 ± 17	−341 ± 13

SCANNING ELECTRON MICROSCOPY (SEM) ANALYSIS OF MICROBIAL MORPHOLOGIES IN THE REACTOR

The sludge in each compartment was observed 40 days after the start-up of the reactor using SEM, as illustrated in Figure 6A–E. The sludge and microscopic results indicated rich biofacies, whereas the dominant cells appeared to have different morphologies in each compartment of the reactor. Different bacterial morphologies were also observed within the reactor. In Compartment 1, the microbial structure developed a heterogeneous colonization pattern composed mainly of cocci and bacilli of varying sizes and shapes, and hydrolyzing and acid-producing microorganisms predominated. There were no predominant bacteria in Compartment 2. As shown in Figure 6B, cocci, bacilli, and filamentous microorganisms adhered to the sludge. Based on their bionomics, the acidogenic bacteria and methanogens could potentially coexist in this compartment. The microorganisms in the latter compartments of the reactor were unitary and primarily cocci, as illustrated in Figure 6C–E. These cocci resembled species of the *Methanococcus* genus. The number of smaller cocci decreased from Compartments 3 to 5, whereas the number of larger cocci increased from Compartments 3 to 5. There were numerous holes in the granular sludge that allowed gas to escape and matter to be removed.

The unique flow characteristics of the anaerobic baffled filter reactor allowed different bacterial populations to dominate in each compartment. Therefore, the microbial morphology differed in each compartment. Acidogenesis and methanogenesis were separated longitudinally, allowing the reactor to behave as a two-phase system without the associated control problems and high costs [29]. In addition, the anaerobic baffled filter reactor promoted microbial enrichment and selection by producing biofilm sludge and granular sludge [30].

Figure 6: Scanning electron microscopy (SEMs) illustrating the different types of bacteria in the granular sludge from Compartments 1 to 5, respectively. (A) Compartment 1; (B) Compartment 2; (C) Compartment 3; (D) Compartment 4; (E) Compatment 5.

Archaeal Denaturing Gradient Gel Electrophoresis (DGGE) Analysis

The DGGE approach has been widely used in environmental microbiology to study microbial community structures in complex systems, including anaerobic bioreactors in the laboratory [31,32,33]. Archaeal community fingerprints from each compartment on the sampling dates (on Days 2 and 100) were generated by DGGE analysis of the V5-6 region of 16S rDNA (Figure 7). Thirteen bands, designated Ad1-13, were purified for subsequent sequencing analyses. The band numbers and intensities of the DGGE community fingerprints changed as the operation of the reactor proceeded. The community structures of the samples from the five compartments on Day 2 were similar because the reactor was allowed to reach the stationary phase for 10 days to stimulate biofilm growth on the soft filler in each compartment. The DGGE graph illustrates that Ad1, Ad5, Ad8, and Ad9 were predominant in each compartment of the ABR filled with soft filler on Day 2. Ad1 and Ad8 were closely related to *Methanoregula* species, with 97.5% and 90.2% similarities, respectively.

The microbial community structure and function changed concomitantly during the reactor's operation, and the biomass had sufficient time to adjust and perform more efficiently [34]. Ad3, Ad4, Ad10, Ad11, Ad12, and Ad13 were detected only on Day 100 in the reactor. Two of these species (Ad3 and Ad11) were closely related to *Methanobacteriaceae*, with approximately 92.1%–95.8% similarities, respectively. Ad7 was closely related to *Methanoculleus*, with 90.2% similarity, and *Methanoculleus* has been previously reported in sludge samples obtained from an EGSB [35].

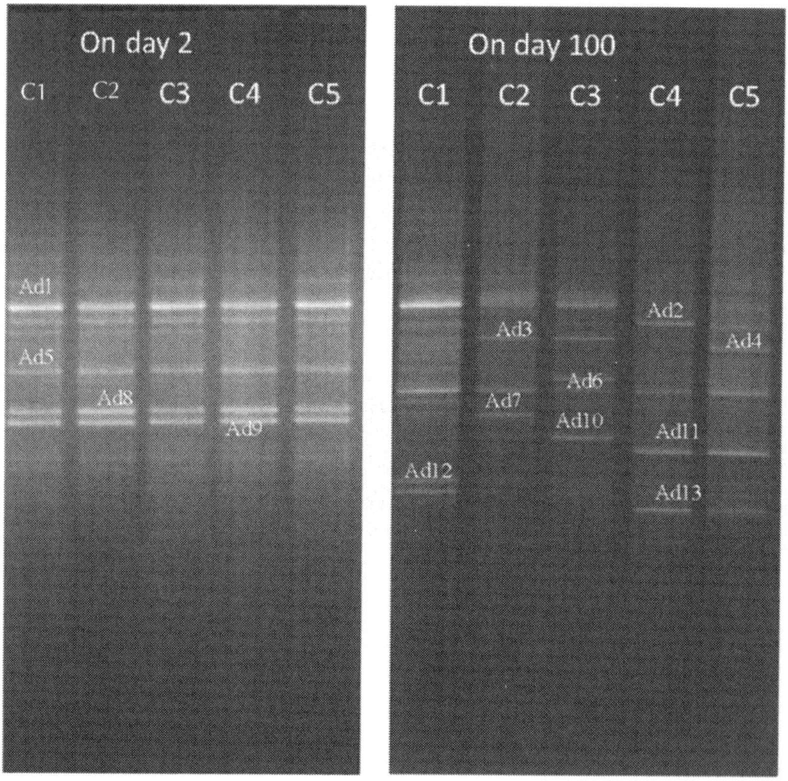

Figure 7: Denaturing gradient gel electrophoresis (DGGE) fingerprints from each compartment of the reactor on Days 2 and 100.

Table 2 presents the phylogenetic affiliations of the band sequences, which were determined by comparison with the Gen Bank database. The DGGE community fingerprints in the samples varied in terms of band number and intensity in each compartment on day 100. The most distinct spatial community succession was detected on day 100, and a clear difference existed between the reactor's anterior and posterior compartments. All of the sequences from the archaeal libraries matched with methanogens of the phylum *Euryarchaeota* and could be classified into five genera: *Methanoregula, Methanobacteriaceae, Methanosaeta, Methanoculleu,* and *Thermogymnomonas*. Ad1 and Ad2 were affiliated with *Methanoregula,* as previously observed in anaerobic sludge [36].

The anterior (Compartments 1 and 2) of the reactor was able remove COD with high efficiency (Figure 2) because microbial organisms with the ability to hydrolyze substances existed in this section of the ABR. Other researchers have observed that some microbial organisms are able to degrade solvents and VFAs through their physiological functions and phylogeny [37,38]. The methanation reaction was the main reaction occurring in the posterior (Compartments 4 and 5) of the ABR, and an abundance of *Methanoregula, Methanobacteriaceae,* and *Thermogymnomonas* (Ad2, Ad11, and Ad13, respectively) was detected in Compartments 4 and 5 on day 100. The anaerobic baffled filter reactor, combined with the biofilm attached to the soft filler and the granular sludge suspended at the bottom of the reactor, promoted microbial selection and zoning. These observations demonstrated that biological phase separation occurred in the ABR. It was unlikely that a complete separation of phases (acidogenic and methanogenic) occurred within the ABR because methanogens were observed in all compartments.

Table 2: Identity of the genomic sequences in the dominant DGGE bands obtained by sequencing and basic local alignment search tool (BLAST) analysis.

Band	Nearest species and taxon	Phylogenetic affiliation to family or genus level	Similarity (%)	Access No.
Ad1	*Candidatus Methanoregulaboonei*	*Methanoregula*	97.5	EU887810
	Methanoregulaformicica	*Methanoregula*	92.1	AB479390
Ad2	*Methanomicrobialesarchaeon*	*Methanoregula*	95.6	AB236088
Ad3	uncultured archaeon	*Methanobacteriaceae*	92.1	EU155919

Ad4	*Candidatus Methanoregulaboonei*	*Methanoregula*	96.5	EU887826
Ad5	uncultured euryarchaeote	*Methanoregula*	90.2	DQ676264
	uncultured archaeon	*Methanoregula*	90.2	EU155924
Ad6	*toluene-degrading methanogenic consortium archaeon*	*Methanosaeta*	95.1	AF423188
	uncultured *Methanosaeta sp.*	*Methanosaeta*	93.2	EU888815
Ad7	uncultured archaeon	Methanoculleus	90.2	FR832414
Ad8	*Methanosaetaconcilii*	*Methanosaeta*	91.6	X16932
Ad9	*Methanosphaera sp.* uncultured bacterium	*Methanobacteriaceae* *Methanobacteriaceae*	94.7 91.7	AF424765 AB266909
Ad10	*Methanosaetaconcilii*	*Methanosaeta*	93.7	X51423
Ad11	*Methanosphaera* sp.	*Methanobacteriaceae*	95.8	AJ576239
Ad12	uncultured Thermoplasmatalesarchaeon	*Thermogymnomonas*	95.1	EU731598
Ad13	uncultured archaeon *Thermoplasmatalesarchaeon*	*Methanoculleus* *Thermogymnomonas*	92 90.1	FR832414 GNA10F01

ENZYME ACTIVITY ANALYSIS

The activities of enzymes, such as those of protease and coenzyme F_{420}, have been extensively studied in anaerobic digestion [14]. These activities play a significant role in the decomposition of substrates, affecting nutrients and nutrient availability. Microbial enzyme activities are associated with microbial metabolism [39]. The coenzyme F_{420} is widely distributed in methanogenia [15].

The results of this study suggest that the activity of the protease in the reactor increased in each compartment over the study period (Figure 8). The protease activity decreased longitudinally along the reactor from Compartments 1 to 5. The highest protease activity was observed in the first compartment, with average values of 27 μmol/(min·g) in Compartment 1, 25 μmol/(min·g) in Compartment 2, 22 μmol/(min·g) in Compartment 3, and less than 3 μmol/(min·g) in Compartments 4 and 5. The protease activity demonstrated that hydrolysis and acidogenesis were the main biochemical reactions occurring in the first few compartments. These observations suggest that the ABR system facilitated systematic selection in the different compartments to promote phase separation.

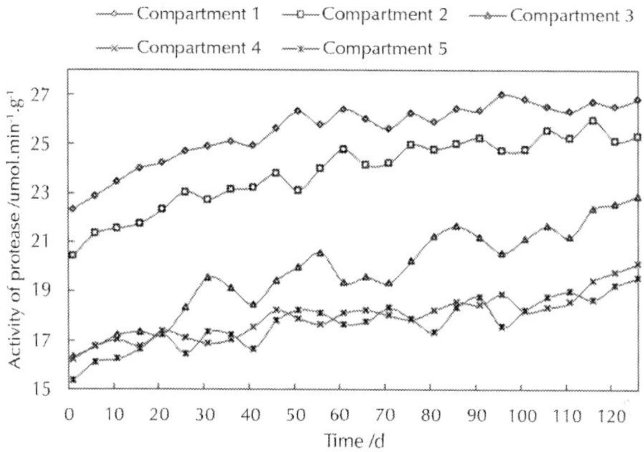

Figure 8: Variation in protease activity in the reactor.

As shown in Figure 9, the activity of coenzyme F_{420} clearly increased during the adaptive transient period of the methanogens after inoculation. The results of the treatment indicate that the coenzyme F_{420} activities in Compartments 3, 4 and 5 were higher than those in Compartments 1 and 2.

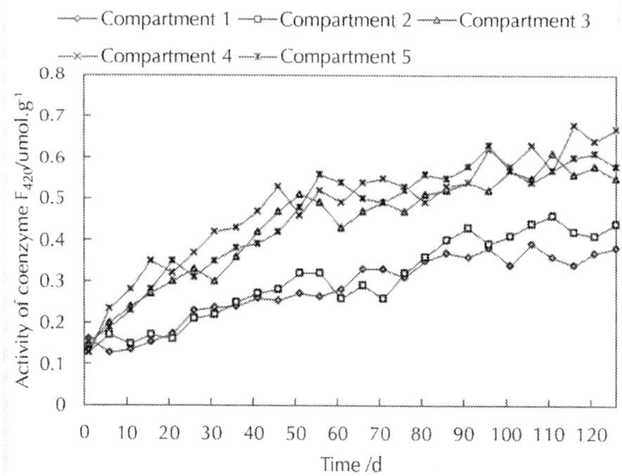

Figure 9: Variation in coenzyme F420 activity in the reactor.

This behavior indicated that methanogenesis was also dominant in the last few compartments. It was unlikely that a complete separation of phases (acidogenic and methanogenic) occurred within the ABR because protease and coenzyme F_{420} were detected in all compartments.

CONCLUSIONS

The main objectives of this study were to examine the treatment of algae-laden water using an anaerobic baffled filter reactor and to elucidate the microbial phase in the system during the experiments. The algae-laden water was treated with an OLR of approximately 1.5 kg COD/(m^3d) at an HRT of five days and an ambient temperature of 30 °C, which resulted in an 80% COD removal efficiency and a 293 mL/(Ld) biogas production rate. The spatiotemporal community succession patterns were altered, and the DGGE community fingerprints in the samples differed with respect to band number and intensity in each compartment on day 100 but were similar on Day 2. Moreover, five dominant sequence genera, *Methanoregula, Methanobacteriaceae, Methanosaeta, Methanoculleu*, and *Thermogymnomonas*, were detected in the sludge and classified by DGGE.

The anaerobic baffled filter reactor, combined with the biofilm sludge attached to the soft filler and granular sludge suspended at the bottom of the reactor, promoted microbial selection and zoning. The protease activity in the reactor decreased longitudinally along the reactor from Compartments 1 to 5, but the activity of coenzyme F_{420} increased. The anaerobic baffled filter reactor developed in this study was concluded to be an efficient reactor configuration for the treatment of algae-laden water. Post-treatment, such as biological contact oxidation and aquatic vegetation wetland use, was required to treat the anaerobic effluent and satisfy Chinese standard GB18918-2002 for the discharge of pollutants.

ACKNOWLEDGMENTS

This work was supported by the National Major Projects of Water Pollution Control and Management Technology of China

(2012ZX07101-005), National Natural Science Foundation of China (51009027) and the research fund of the Key Laboratory for Advanced Technology in Environmental Protection of Jiangsu Province.

CONFLICTS OF INTEREST

The authors declare no conflict of interest.

REFERENCES

1. Eleuterio, L.; Batist, J.R. Biodegradation studies and sequencing of microcystin-LR degrading bacteria isolated from a drinking water biofilter and a fresh water lake. *Toxicon* 2010, *55*, 1434–1442.
2. Guo, L. Doing battle with the green monster of Taihu lake. *Science* 2007, *317*.
3. Yan, Q.; Zhao, M.; Miao, H.; Ruan, W.; Song, R. Coupling of the hydrogen and polyhydroxyalkano-ates (PHA) production through anaerobic digestion from Taihu blue algae. *Bioresour. Technol.* 2010, *101*, 4508–4512.
4. Cai, Y.; Kong, F.; Shi, L.; Yu, Y. Spatial heterogeneity of cyanobacterial communities and genetic variation of Microcystis populations within large, shallow eutrophic lakes (Lake Taihu and Lake Chaohu, China). *J. Environ. Sci.* 2012, *24*, 1832–1842.
5. Liu, Y.; Chen, W.; Li, D.; Huang, Z.; Shen, Y.; Liu, Y. Cyanobacteria-cyanotoxin-contaminations and eutrophication status before Wuxi Drinking Water Crisis in Lake Taihu, China. *J. Environ. Sci.* 2011, *23*, 575–581.
6. Wu, B.; Zhao, D.; Zhang, Y.; Zhang, X.; Cheng, S. Multivariate statistical study of organic pollutants in Nanjing reach of Yangtze River. *J. Hazard. Mater.* 2009, *169*, 1093–1098.
7. Kuscu, O.S.; Sponza, D.T. Performance of anaerobic baffled reactor (ABR) treating synthetic wastewater containing p-nitrophenol. *Enzym. Microb. Technol.* 2005, *36*, 888–895.
8. Uyanik, S.; Sallis, P.J.; Anderson, G.K. The effect of polymer addition on granulation in an anaerobic baffled reactor (ABR). Part 1: Process performance. *Water Res.* 2002, *36*, 933–943.

9. Ji, R.P.; Lu, X.W.; Li, X.N.; Pu, Y.P. Biological degradation of algae and microcystins by microbial enrichment on artrificial media. *Ecol. Eng.* 2009, *35*, 1584–1588.
10. Bachmann, A.; Beard, V.L.; McCarty, P.L. Performance characteristics of the anaerobic baffled reactor. *Water Res.* 1985, *19*, 99–106.
11. State Environmental Protection Administration of China. *National Standard Methods for Water and Wastewater Quality Analysis*, 4th ed.; China Environmental Science Press: Beijing, China, 2004; pp. 102–226.
12. Liu, R.; Lu, X.; Tian, Q.; Yang, B.; Chen, J. The performance evaluation of hybrid anaerobic baffled reactor for treatment of PVA-containing desizing wastewater. *Desalination* 2011, *271*, 287–294.
13. Zhu, G.-F.; Li, J.-Z.; Wu, P.; Jin, H.-Z.; Wang, Z. The performance and phase separated characteristics of an anaerobic baffled reactor treating soybean protein processing wastewater. *Bioresour. Technol.* 2008, *99*, 8027–8033.
14. Whiteley, C.G.; Heron, P. The enzymology of sludge solubilisation utilizing sulphate reducing systems properties of proteases and phosphatases. *Enzym. Microb. Technol.* 2001, *31*, 419–424.
15. Whitmore, T.N.; Etheridge, S.P.; Stafford, D.A.; Leroff, U.E.A.; Hughes, D. The evaluation of anaerobic digester performance by coenzyme F_{420} analysis. *Biomass* 1986, *9*, 29–35.
16. Arooj, M.F.; Han, S.-K.; Kim, S.-H.; Kim, D.-H.; Shin, H.-S. Sludge characteristics in anaerobic SBR system producing hydrogen gas. *Water Res.* 2007, *41*, 1177–1184.
17. Mohan, S.V.; Raghavulu, S.V.; Goud, R.K.; Srikanth, S.; Babu, V.L.; Sarma, P.N. Microbial diversity analysis of long term operated biofilm configured anaerobic reactor producing biohydrogen from wastewater under diverse conditions. *Int. J. Hydrog. Energy* 2010, *35*, 12208–12215.
18. Yu, Y.; Lü, X.; Li, J. Start-up of improved external circulation anaerobic reactor treating cyanobacteria and microbial properties of granular sludge. *CIESC J.* 2013, *64*, 4203–4219.
19. Du, J.; Yan, S.-H.; Chang, Z.-Z. Potential of methane production of blue algae and its feasibility of fermentation with anaerobic baffled reactor. *Jiangsu Agric. Sci.* 2008, *24*, 948–953.

20. Yu, Y.; Lü, X.; Wu, Y.; Xu, W.; Zhao, J.-W. Research on combined process of biological contact oxidation and aquatic vegetable wetland for algae anaerobically digested effluent treatment. *J. Cent. South Univ. Nat. Sci.* 2013, *40*, 87–92.
21. Uyanik, S.; Sallis, P.J.; Anderson, G.K. The effect of polymer addition on granulation in an anaerobic baffled reactor (ABR). Part II: Compartmentalization of bacterial papulations. *Water Res.* 2002, *36*, 944–955. [Google Scholar] [CrossRef]
22. Holst, T.; Jorgensen, N.O.; Jorgensen, C.; Johansen, A. Degradation of microcystin in sediments at oxic and anoxic, denitrifying conditions. *Water Res.* 2003, *37*, 4748–4760.
23. Falconer, I.R.; Humpage, A.R. Health risk assessment of cyanobacterial (blue-green algal) toxins in drinking water. *Int. J. Environ. Res. Publ. Health* 2005, *2*, 43–50.
24. Nasri, H.; El Herry, S.; Bouaicha, N. First reported case of turtle deaths during a toxic Microcystis spp. bloom in Lake Oubeira, Algeria. *Ecotoxicol. Environ. Safe* 2008, *71*, 535–544.
25. Mensah, K.A.; Forster, C.F. An examination of the effects of detergents on anaerobic digestion. *Bioresour. Technol.* 2003, *90*, 133–138.
26. Ren, N.Q.; Zhao, D.; Chen, X.L.; Li, J.Z. Mechanism and controlling strategy of the production and accumulation of propionic acid for anaerobic wastewater treatment. *Sci. China B* 2002, *45*, 319–327.
27. Speece, R.E. *Anaerobic Biotechnology for Industrial Wastewater*; Archae Press: Nashville, TN, USA, 1996; pp. 45–49. [Google Scholar]
28. Dolfing, J.; Bloeman, W.G.B.M. Activity measurements as a tool to characterize the microbial composition of methanogenic environments. *J. Microbiol. Methods* 1985, *4*, 1–12.
29. Barber, W.P.; Stuckey, D.C. The use of the anaerobic baffled reactor (ABR) for wastewater treatment: A review. *Water Res.* 1999, *33*, 1559–1578.
30. Angenent, L.T.; Abel, S.J.; Sung, S. Effect of an organic shock load on the stability of an anaerobic migrating blanket reactor. *J. Environ. Eng.* 2002, *128*, 1109–1120.

31. Lee, C.; Kim, J.; Shin, S.G.; Hwang, S. Monitoring bacterial and archaeal community shifts in a mesophilic anaerobic batch reactor treating a high-strength organic wastewater. *FEMS Microbiol. Ecol.* 2008, *65*, 544–554.
32. Diaz, E.E.; Stams, A.J.A.; Amils, R.; Sanz, J.L. Phenotypic properties and microbial diversity of methanogenic granules from a full-scale upflow anaerobic sludge bed reactor treating brewery wastewater. *Appl. Environ. Microbiol.* 2006, *72*, 4942–4949.
33. Bialek, K.; Kim, J.; Lee, C.; Collins, G.; Mahony, T.; O'Flaherty, V. Quantitative and qualitative analyses of methanogenic community development in high-rate anaerobic bioreactors. *Water Res.* 2011, *45*, 1298–1308.
34. Liu, X.; Ren, N.; Yuan, Y. Perforrnance of a periodic anaerobic baffled reactor fed on Chinese traditional medicine industrial wastewater. *Bioresour. Technol.* 2009, *100*, 104–110.
35. Collins, G.; Woods, A.; McHugh, S.; Carton, M.W.; OFlaherty, V. Microbial community structure and methanogenic activity during start-up of psychrophilic anaerobic digesters treating synthetic industrial wastewaters. *FEMS Microbiol. Ecol.* 2003, *46*, 159–170.
36. Watanabe, R.; Tada, C.; Baba, Y.; Fukuda, Y.; Nakai, Y. Enhancing methane production during the anaerobic digestion of crude glycerol using Japanese cedar charcoal. *Bioresour. Technol.* 2013, *150*, 387–392.
37. Zhang, J.; Wei, Y.; Xiao, W.; Zhou, Z.; Yan, X. Performance and spatial community succession of an anaerobic baffled reactor treating acetone-butanol-ethanol fermentation wastewater. *Bioresour. Technol.* 2011, *102*, 7407–7414.
38. Nachaiyasit, S.; Stuckey, D.C. The effect of shock loads on the performance of an anaerobic baffled reactor (ABR). 2. Step and transient hydraulic shocks at constant feed strength. *Water Res.* 1997, *31*, 2747–2754.
39. Aoki, K.; Miyamoto, K.; Murakami, S. Anaerobic synthesis of extracellular proteases by the soil bacterium bacillus sp.AM-23: Purification and characterization of the enzymes. *Soil Biol. Biochem.* 1995, *27*, 1377–1382.

Chapter 8

Monitoring Urban Wastewaters' Characteristics by Visible and Short Wave Near-Infrared Spectroscopy

Ignacio Melendez-Pastor, M. Belén Almendro-Candel, Jose Navarro-Pedreño , Ignacio Gómez , Mariano García Lillo and Encarni I. Hernández;

Department of Agrochemistry and Environment, University Miguel Hernández, Av Universidad s/n Edificio Alcudia, Elche 03202, Spain;

ABSTRACT

On-line monitoring of wastewater parameters is a major scientific and technical challenge because of the great variability of wastewater characteristics and the extreme physical-chemical conditions that endure the sensors. Wastewater treatment plant managers require fast and reliable information about the input sewage and the operation of the different treatment stages. There is a great need for the development of sensors for the continuous monitoring of wastewater parameters. In this sense, several optical systems have been evaluated. This article presents an experimental laboratory-based approach to quantify commonly employed urban wastewater parameters, namely biochemical oxygen demand in five days (BOD_5), chemical oxygen demand (COD), total suspended solids (TSS), and the ratio BOD_5:COD, with a visible and short wave near infrared (V/SW-NIR) spectrometer (400–1000 nm). Partial least square regression (PLSR) models were developed in order to quantify the wastewater parameters with the recorded spectra. PLSR models were developed for the full spectral range and also for the visible and near infrared spectral ranges separately. Good PLSR models were obtained with the visible spectral range for BOD_5 (RER = 9.64), COD (RER = 10.88), and with the full spectral range for the TSS (RER = 9.67). The results of this study show that V/SW-NIR spectroscopy is a suitable technique for on-line monitoring of wastewater parameters.

INTRODUCTION

In recent decades, a large number of wastewater treatment plants (WWTP) has been constructed in response to environmental demands by the societies and the legislation [1]. Conventional WWTP comprise a series of treatment stages designed to treat large volumes of wastewater to acceptable regulatory standards. Most of the WWTP include a succession of unitary operations, for example, sedimentation tanks, coagulation-flocculation chambers, biological reactors or even final disinfection systems. The complex succession of unitary operations needs to be sharply operated in order to achieve an effective reduction of pollution. In this sense, the most challenging issue is the great variability of wastewater volume and contamination level of the incoming sewage.

The operation of a WWTP needs the continuous monitoring of contamination level of incoming wastewater and the effective pollution reduction along each stage of the treatment process. On-line monitoring of wastewater quality parameters through the treatment process is a major technical challenge because of the spatial and time dependent variability of wastewater characteristics [2]. Some traditional wastewater quality parameters include the Biochemical Oxygen Demand within five days (BOD_5), Chemical Oxygen Demand (COD) and Total Suspended Solids (TSS). The quantification of these parameters is time consuming and implies a notable continuum cost. A fast, accurate and cost-effective system to continuously monitor wastewater parameters through the treatment process may allow operators to optimize the hydraulic retention times (and associated electrical energy cost) and the doses of reactants for some of the stages (e.g., coagulant-flocculent).

The employment of optical-electronic systems for the online monitoring of wastewater treatment process is a relatively recent approach [3] but with a promising future [2,4]. Many studies have employed optical-electronic systems to quantify the relationships among several wastewater parameters and several UV-VNIR (ultraviolet—visible and near infrared) spectral features. Thomas *et al.* [5] employed an UV detector to estimate total organic carbon (TOC), TSS, COD and BOD. Great effort has been focused in the application of fluorescence spectroscopy to improve the quantification of BOD_5 [6,7]. In this sense, the application of UV spectroscopy for monitoring wastewater parameters has been greatly improved and new statistical methods, such as neural networks [8], are currently being researched. Moreover, the application of VNIR spectroscopy for wastewater quality is an area of current research with great challenges ahead but with a great potential for on-line monitoring of wastewater parameters [9,10].

Spectroscopic analysis of complex media, such as wastewater, makes the task of implementing optical-electronic systems for the online monitoring of wastewater treatment more difficult. Thus, researchers are encouraged to exploit the full potential of spectral information, to analyze the absorbance that describes the chemistry and the scatter that is related with the particles size and distribution in complex dispersive media [11]. The scatter is generally considered as a "parasitic" phenomenon, complicating the spectroscopic analysis of complex dispersive media, but the detected intensities of scattered light at different wavelengths depend on the number and sizes of colloidal particles, and consequently, on the respective component content [12]. Scattering in the visible and near infrared tends to be lower than for the UV spectral range (*i.e.*, scattering is inversely proportional to the wavelength). The effect of suspended particles in wastewater samples for UV spectroscopy systems without an integrating sphere assembly that take into account for scattered UV light, is the over-prediction of absorbance [13,14] and lower predictions accuracy. Scatter-based spectroscopic analysis of multi-component mixtures in the visible and short wave near infrared (SW-NIR) has been previously studied in several industrial applications, thus providing suitable quantitative information of the studies [11,12,15,16,17].

This study focused on the use of visible and short wave near infrared (V/SW-NIR) spectroscopy to quantify wastewater quality parameters (*i.e.*, BOD_5, COD, and TSS) in samples from an urban WWTP, collected at different treatment stages (*i.e.*, input sewage, after a physical-chemical treatment, and after a biological treatment).

MATERIALS AND METHODS

Wastewater samples were collected at an urban WWTP serving Alicante in the Southeast of Spain during the spring season. Samples were obtained along four months and at different weekdays in order to obtain a larger variability of sewage inputs. Three different locations within the treatment processes were sampled: (1) at the entrance to the WWTP, where water continuously flowed through a set of sieves with a minimum aperture of around 1 cm; (2) after the primary treatment, consisting of a series of aeration tanks for sand-fat removal and coagulation-flocculation chamber and subsequent decantation; (3) after the biological treatment, consisting of a biological reactor and subsequent decantation. A total of 84 samples were obtained (*i.e.*, 28 samples per treatment stage) and analyzed in this experiment. Water samples were collected and immediately stored under cold conditions (~4 °C) to minimize water degradation.

Water analyses were conducted within hours from the sampling (less than 4 hours). Selected wastewater quality parameters were the biochemical oxygen demand within five days (BOD_5), chemical oxygen demand (COD), and total suspended solids (TSS) that were analyzed with standard laboratory methods. BOD5 was determined after 5-days of incubation at dark constant-temperature (20 ± 1 °C) conditions. Water samples were placed in incubation bottles on agitation racks, and included a manometer to quantify oxygen consumption. Chemical oxygen demand was determined with a closed reflux, colorimetric method, and total suspended solids were determined by gravimetry of the increase of weight by the residue retained on a filter dried to a constant weight at 103–105 °C [18].

These parameters are the existing standard and most widely applied methods for organic load monitoring [2]. Also, the ratio BOD_5:COD was computed as a proxy of wastewater biodegradability. As a reference value, an ideal biodegradability index is a BOD_5:COD ratio close to 1.0 [19]. Four replicates per water sample were done for all wastewater parameters and the average values of the replicates were used for further analyses.

Spectral Measurements

An ASD Field Spec Hand Held VNIR radiometer (Analytical Spectral Devices Inc., Boulder, CO, USA) was utilized to measure wastewater spectra. This radiometer covers the wavelength range of 325–1075 nm, which approximates with the visible (V) and short wave near infrared (SW-NIR) spectral regions, an accuracy of ±1 nm and a resolution of <3 nm at 700 nm. The radiometer was connected through a fiber optic cable to an Ocean Optics (Leesburg, FL, USA) cuvette holder where water samples were placed. Another fiber optic cable was connected from the cuvette holder to an ASD Fiber Optic Illuminator® as light source. This systems enables the illumination of a 10 mm cuvette from one of its faces and the transmittance record from the opposite cuvette face.

Radiometric measurements were also conducted within hours from the sampling in order to minimize the alteration of the wastewater samples. All spectra were acquired in transmittance mode using distilled water as blank. Five radiometric measurements (with 15 automatic replicate spectra per measurement) were taken for each water sample. The dark current (detector background) and reference spectra were taken immediately before each spectral measurement. The five radiometric measurements were visually inspected and then averaged to obtain a single spectrum per water sample (n = 84). Random noise was minimized by applying a Savitzky-Golay algorithm across a moving window of 10 nm with a 3rd order polynomial [20]. Spectra were acquired at room temperature (24 ± 1 °C). All samples were homogenized (15 s) by mechanical mixing, and then immediately placed in the cuvette. Homogenization is a useful sample pretreatment, because it does not remove the scatter but simplifies its effect on spectra, thus, facilitating the data analysis [11].

Spectroscopic Analyses

Partial Least Squares Regression (PLSR) was the selected statistical technique to relate the wastewater quality parameters with the water spectra. PLSR has been designed to confront the situation that there are many, possibly correlated, predictor variables, and relatively few samples [21]. In this sense, PLSR provides feasible quantitative multivariate modeling methods for chemometrics [22] where highly detailed spectra data (*i.e.*, high spectral resolution or number of bands) are employed to quantitatively predict a limited number of problem samples. In this study, PLSR models were developed for a sensor-noise-free spectral range. Models were developed for the wavelength range 400–1000 nm and also for two smaller spectral ranges of 300 nm each one. These spectral subsets were identified with the visible (400–700 nm) and the near infrared (700–1000 nm) spectral ranges [23]. PLSR models and further statistical analyses were developed with the R statistical programming language [24].

The selection and evaluation process of the PLSR models [21] was based on the following methodological procedure: (1) the original 84 samples dataset was randomly divided into an initial 60 samples dataset for model cross-validation (with 20 samples per treatment stage), and the 24 remaining samples for independent test; (2) a leave-one-out (LOO) cross-validation (CV) procedure was used for the development of PLSR models with a dataset of 60 samples [25]; and (3) selected models were tested with the independent validation dataset in order to assess the predictive capabilities of the selected PLSR models.

Several diagnostic statistics were employed for PLSR models assessment. Cross-validation Pearson correlation coefficient (R^2) was employed as an illustrative diagnostic statistic. The root mean squared error (RMSE) was the fundamental statistical parameter used to guide the number of model components or latent variables (LV) selection. RMSE is calculated as:

$$\mathrm{RMSE} = \sqrt{\frac{1}{N}\sum_{i=1}^{N}\left(Z(Xi) - Z^{*}(Xi)\right)^{2}} \tag{1}$$

Where N is the sample size, $Z(x_i)$ is the observed value at location i and $Z^*(x_i)$ is the predicted value at location i. In addition, the bivariate RMSE is suitable for overall measurements of model performance [26]. The number of optimal components was determined based on the lowest RMSE values for the adjusted CV with the LOO procedure (*i.e.*, RMSECV). In addition, the range error ratio (RER) was used to determine the practical utility of the models [27]. The RER is computed as:

$$\mathrm{RER} = \frac{\mathrm{Range}}{\mathrm{RMSEP}} \tag{2}$$

Where the numerator is the range of the dataset and RMSEP in denominator is the model error of prediction. The RER was employed in the prediction test stage in order to compare the practical utility of the spectroscopic approach to predict the different wastewater parameters.

RESULTS

Several descriptive statistical parameters of the wastewater parameters were computed (Table 1). A great reduction of the magnitude of wastewater parameters was observed along the wastewater treatment process. Average BOD_5 was reduced from 461.0 mg/L of the raw sewage, to 202.1 mg/L and 17.7 mg/L after the primary and secondary treatments respectively. Similarly, average COD was reduced from 946.1 mg/L of the raw sewage, to 407.4 mg/L and 57.6 mg/L after the primary and secondary treatments respectively. The TSS greatly reduced from the raw sewage (471.8 mg/L) to the primary treatment outlet (131.0 mg/L) and also after the secondary treatment (19.3 mg/L). The ratio BOD_5:COD changed from about 0.5 for the raw sewage (0.483) and primary treatment (0.502), to 0.307 for the secondary treatment effluent. All of these parameters provide an overview of the main characteristics of the wastewaters analyzed.

The characteristic spectra for the three different wastewater treatment stages are shown (Figure 1). The relative transmittance spectra were highly different according to the wastewater treatment stage. The wastewater treatment process promoted a notable increase of the relative transmittance of the samples, especially for the near infrared spectral range. Raw sewage samples were dark brown with low relative transmittance values, ranging from 0.03 at 400 nm to 0.26 at 1000 nm. Primary treatment relative transmittance ranged from 0.05 at 400 nm to 0.49 at 1000 nm. Secondary treatment relative transmittance ranged from 0.66 at 400 nm to 0.93 at 1000 nm. Secondary treatment promoted the most accentuated clarification of the wastewater samples, as observed by the greater relative transmittance increment as compared with the primary treatment samples. No other accentuated characteristic spectral features were observed.

Table 1: Summary statistics of the calibration dataset for Biochemical Oxygen Demand within five days (BOD5), Chemical Oxygen Demand (COD), Total Suspended Solids (TSS), and BOD_5: COD at selected treatment stages.

Variables	Statistics	Treatment stage		
		Raw sewage	Primary	Secondary
BOD_5 (mg/L)	Mean	461.0	202.1	17.7
	St. Dev.	147.3	46.9	8.1
	Minimum	280	120	6
	Maximum	920	260	34
COD (mg/L)	Mean	946.1	407.4	57.6
	St. Dev.	242.6	106.2	13.6
	Minimum	615	197	35
	Maximum	1522	572	83
TSS (mg/L)	Mean	471.8	131	19.3
	St. Dev.	154.8	27.9	8.5
	Minimum	290	79	6
	Maximum	902	172	35

BOD_5:COD	Mean	0.483	0.502	0.307
	St. Dev.	0.048	0.041	0.076
	Minimum	0.437	0.451	0.178
	Maximum	0.613	0.609	0.410

PLSR Models Calibration and Validation

The number of latent variables or model components was selected based on the minimization of the RMSECV. The number of components ranged from 2 to 4 for all variables (Table 2). The Pearson coefficient (R^2) was also computed for a general overview of the models performance. Pearson coefficient values ranged from about 0.87 for BOD_5, COD, and TSS, to about 0.6 for the BOD_5: COD ratio. RMSECV values for BOD_5 were about 78 mg/L that is about 8.5% of the range of the BOD_5 calibration values. The absolute values of the RMSECV for COD were slightly higher (~139 mg/L), but were very similar to the previous variable as expressed relatively to the range of the COD calibration values (9%). TSS calibration results were also very similar to the previous variables, with a RMSECV about 79 mg/L (8.8% of the range). Calibration results for the BOD_5:COD ratio reported a RMSECV for the absolute value about 0.06, with an RMSECV value about 14%–15% relative to the range of the variable. The best PLSR were obtained for the visible spectral range (400–700 nm) for all wastewater parameters with the exception of the TSS that performed better for the full spectral range (400–1000 nm).

Models selected from the cross-validation stage were employed with the independent test dataset in order to assess their predictive capability and generalization (Table 2). The magnitude of the RMSE values of the predictions (RMSEP) was similar to the RMSECV. RMSEP for the BOD_5 independent validation dataset was 77.81 (10.37%) and RER was 9.64 for the visible spectral range (400–700 nm). RMSEP for the COD independent validation dataset was 128.40 (9.19%) with a RER of 10.88 for the visible spectral range (400–700 nm). RMSEP for the TSS independent validation dataset was 83.26 (10.34%) with a RER of 9.67 for the full spectral range (400–1000 nm). RMSEP for the BOD_5:COD independent validation dataset was 0.059 (17.90%) with a RER of 5.59 for the visible spectral range (400–700 nm).

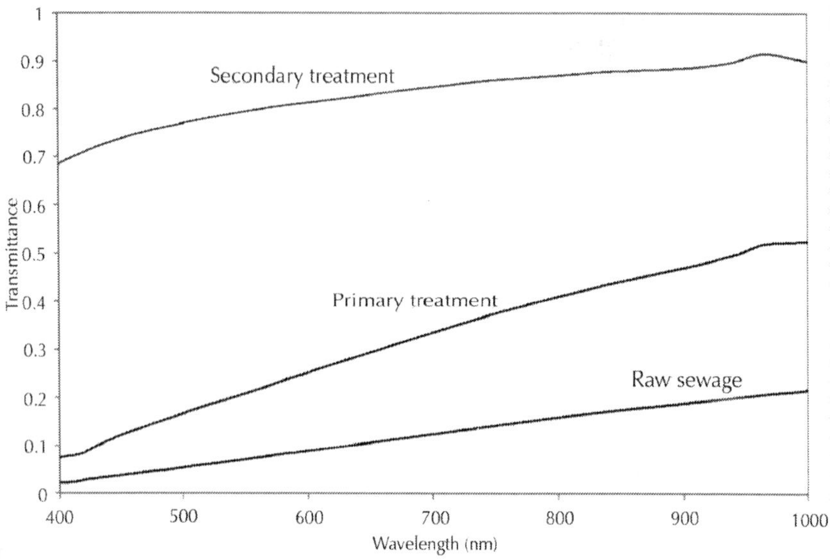

Figure 1: Characteristic relative transmittance spectra for the raw sewage, primary treatment and secondary treatment wastewater samples.

Table 2: Results of the partial least square regression (PLSR) models cross-validation (CV) and prediction (P) models for the BOD5; COD; TSS and the ratio BOD_5:COD wastewater parameters.

Parameters	Spectral range (nm)	LV	Cross-validation			Prediction test		
			R^2	RMSECV	RMSECV (%)	RMSEP	RMSEP (%)	RER
BOD_5	400–1000	2	0.849	78.81	8.62	-		
	400–700	3	0.854	77.53	8.48	77.81	10.37	9.64
	700–1000	2	0.849	78.81	8.62	-		
COD	400–1000	2	0.877	139.1	9.35	-		
	400–700	3	0.878	138.5	9.31	128.40	9.19	10.88
	700–1000	2	0.877	139.4	9.37	-		
TSS	400–1000	4	0.864	78.00	8.71	83.26	10.34	9.67
	400–700	3	0.857	80.09	8.94	-		
	700–1000	3	0.861	79.00	8.82	-		
BOD_5:COD	400–1000	3	0.643	0.062	14.25	-		
	400–700	3	0.644	0.062	14.25	0.059	17.90	5.59
	700–1000	2	0.599	0.066	15.17	-		

DISCUSSION

One major limitation of VNIR spectroscopy is the reduced or absent number of diagnostics bands for chemometric analysis [28]. This implies that single band regression or derivative analyses are really difficult to establish by the absence of accentuated characteristic spectral features. With this premise in mind, this study applied wide spectral ranges to quantify the wastewater parameters based on the general shape or "area under the curve" of the relative transmittance spectra. The effect of the wastewater parameters variability on spectra can be described as spectrum offset and slope differences (Figure 1). This effect has been previously described in complex dispersive media in the visible and short wave NIR by the multiple scattering by the presence of particles with different composition, size and distribution [11,17]. This method is based on an indirect scatter effect and its practical applicability scope depends on the model robustness in the whole range of natural sample variability [12]. In this sense, previous studies have been able to successfully quantify the composition of wastewater sludge with a transflectance probe for the spectral range 900–1700 nm [29].

PLSR allows a feasible modeling of the different wastewater parameters with VNIR spectroscopy by taking into account the full spectral range of the models instead of single bands. The main advantage of PLSR is that it can be used with any number of explanatory variables, generally providing regression models with highest predictive ability with the smallest number of factors as compared with other regression methods such as ordinary least squares estimator or ridge regression [30]. The PLSR modeling approach allowed the development of a full cross-validation scheme with a moderate number of samples and large number of relative transmittance records (*i.e.*, 600 records for the full spectrum). The number of samples in this study (60 cross-validation + 20 prediction) was in accordance to previous studies such as Thomas *et al* [5] that used 86 samples, Reynolds and Ahmad [6] analyzing up to 54 samples from the same wastewater treatment plant, or even higher than other studies than employed around 40 samples [8,9].

PLSR models selection was based on the lowest RMSECV values (Table 2). Many authors reported their model performance solely based on the Pearson correlation coefficient; however, several computed statistics (*i.e.*, RMSE, RER, and R^2) were applied as a guide of this discussion. Additionally, many studies have been based on the UV spectroscopy instead of VNIR spectroscopy but could be valuable to compare the model performance with different optical-electronic systems. Our best BOD_5 model was determined with three latent variables for the visible spectral range (400–700 nm) and a RMSECV of 8.48% (R^2 = 0.854). This correlation coefficient was slightly better than the reported by Thomas *et al.* [5] with an UV system (R^2 = 0.73). Reynolds and Ahmad [6] also employed an UV system for BOD_5 spectroscopy and reported a slightly better correlation coefficient (R^2 ~ 0.94). They concluded that a linear relationship between the BOD_5 of wastewater and their corresponding fluorescence intensities at 340nm exists using an excitation wavelength of 280 nm.

The best COD model was also obtained with three latent variables for the visible spectral range (400–700 nm) and a RMSECV of 9.31% (R^2 = 0.878). Although the RMSECV of the COD was slightly higher than for the BOD_5, the correlation coefficient was better as previously reported by Thomas *et al.* [5] that obtained a R^2 of 0.92 for the chemical oxygen demand instead of the previously noted 0.73 for the biochemical oxygen demand. Fogelman *et al.* [8] predicted COD of wastewater with an artificial neural network for the spectral range 190–350nm. They reported a correlation coefficient R = 0.96 (*i.e.*, R^2 = 0.92), which is slightly better than our visible spectral range PLSR model. They also noted that the addition of covariates, like the turbidity information, did not significantly improve the accuracy of the artificial neural network. Sarraguça *et al.* [9] monitored COD of an activated sludge reactor with an UV-visible spectrometer with the spectral range 230–700 nm and a NIR spectrometer with the spectral range 900–1700 nm. They also employed PLSR analysis and reported a low correlation coefficient (0.28) with the NIR system and comparable correlation coefficient results (0.82) for the UV-visible system.

The best TSS model was for the full spectral range (400–1000 nm) with four latent variables for a RMSECV of 8.71% (R^2 = 0.864). Thomas *et al.* [5] reported a very similar correlation coefficient (R^2 = 0.87) for an UV system. Sarraguça *et al.* [9] also quantified TSS with their dual UV-visible and NIR spectrometers systems. They reported a good correlation coefficient (0.82) for the UV-visible system and even better for the NIR system (0.92). Their results indicates that TSS was better predicted with greater wavelengths, thus explaining our better performance of the full spectral range for TSS quantification instead of only the visible spectral range reported for the other wastewater parameters. The best BOD_5: COD ratio model was for the visible spectral range (400–700 nm) with three latent variables for a RMSECV of 14.25% ($R^2 = 0.644$). Although the performance of the BOD_5: COD ratio models is lower than for the other parameters, the novelty of this wastewater parameter quantification is a valuable topic for further research and for controlling operation techniques.

RER provided information about the practical utility of the PLSR models for the different wastewater parameters in order to determine which parameters were better modeled. As a general guideline, RER values of between 3 and 10 indicate limited to good practical utility, and values above 10 show that the model has a high utility value [27]. Range error ratio value for the COD parameter was higher than 10 (RER = 10.88) that indicates the capability of our visible and short wave near infrared spectroscopic modeling approach for monitoring chemical oxygen demand in wastewaters. Additionally, RER values very close to 10 were obtained for the BOD_5 (RER = 9.64) and the TSS (RER = 9.67). The modeling results for both parameters were also very promising for the online monitoring. A lower RER value (RER = 5.59) was obtained for the BOD_5:COD ratio.

CONCLUSIONS

This study employed visible and near infrared spectroscopy to quantify wastewater parameters, such as BOD_5, COD, TSS and the BOD_5:COD ratio. Partial least square regression modeling allowed the development of models suitable to quantify wastewater parameters from relative transmittance spectra. The better spectral ranges for wastewater parameters quantification were the visible spectral range (400–700 nm) for BOD_5, COD, and BOD_5: COD ratio, while the TSS were better predicted with the full spectral range (400–1000 nm). The performance of our models was similar to previous studies that used UV fluorescence or UV-VNIR spectrometers. This study provides some valuable information to promote the implementation of VNIR systems for online wastewater monitoring at wastewater treatment plants as a fast and feasible way.

CONFLICTS OF INTEREST

The authors declare no conflict of interest.

REFERENCES

1. European Union. Council Directive 91/271/EEC of 21 May 1991 concerning urban waste-water treatment. pp. L135/40–L135/52. Available online: http://eur-lex.europa.eu (accessed on 8 October 2013).
2. Bourgeois, W.; Burgess, J.E.; Stuetz, R.M. On-line monitoring of wastewater quality: A review. *J. Chem. Technol. Biotechnol.* 2001, *76*, 337–348.
3. Dobbs, R.A.; Wise, R.H.; Dean, R.B. The use of ultra-violet absorbance for monitoring the total organic carbon content of water and wastewater. *Water Res.* 1972, *6*, 1173–1180.
4. Reynolds, D. Prospects for "star wars" monitoring of water and wastewater quality. *Chem. Bus.* 1999, *13*, 113.
5. Thomas, O.; Théraulaz, F.; Cerdà, V.; Constant, D.; Quevauviller, P. Wastewater quality monitoring. *Trends Anal. Chem.* 1996, *16*, 419–424.
6. Reynolds, D.M.; Ahmad, S.R. Rapid and direct determination of wastewater BOD values using a fluorescence technique. *Water Res.* 1997, *31*, 2012–2018.
7. Ahmad, S.R.; Reynolds, D.M. Monitoring of water quality using fluorescence technique: Prospect of on-line process control. *Water Res.* 1999, *33*, 2069–2074.
8. Fogelman, S.; Zhao, H.; Blumenstein, M. A rapid analytical method for predicting the oxygen demand of wastewater. *Anal. Bioanal. Chem.* 2006, *386*, 1773–1779.
9. Sarraguça, M.C.; Paulo, A.; Alves, M.M.; Dias, A.M.A.; Lopes, A.J.; Ferreira, E.C. Quantitative monitoring of an activated sludge reactor using on-line UV-visible and near-infrared spectroscopy. *Anal. Bioanal. Chem.* 2009, *395*, 1159–1166.
10. Ward, A.J.; Hobbs, P.J.; Holliman, P.J.; Jones, D.L. Evaluation of near infrared spectroscopy and software sensor methods for

determination of total alkalinity in anaerobic digesters. *Bioresour. Technol.* 2011, *102*, 4083–4090.

11. Bogomolov, A.; Melenteva, A. Scatter-based quantitative spectroscopic analysis of milk fat and total protein in the region 400–1100 nm in the presence of fat globule size variability. *Chemom. Intell. Lab. Syst.* 2013, *126*, 129–139.
12. Bogomolov, A.; Dietrich, S.; Boldrini, B.; Kessler, R.W. Quantitative determination of fat and total protein in milk based on visible light scatter. *Food Chem.* 2012, *134*, 412–418.
13. Mamane, H.; Ducoste, J.J.; Linden, K.G. Effect of particles on ultraviolet light penetration in natural and engineered systems. *Appl. Opt.* 2006, *45*, 1844–1856.
14. Cantwell, R.; Hofmann, R. Errors in ultraviolet fluence calculations due to particles in wastewater samples. *Environ. Eng. Sci.* 2011, *28*, 159–162.
15. Wu, D.; He, Y.; Feng, S. Short-wave near-infrared spectroscopy analysis of major compounds in milk powder and wavelength assignment. *Anal. Chim. Acta* 2008, *610*, 232–242.
16. Tsenkova, R.; Meilina, H.; Kuroki, S.; Burns, D. Near infrared spectroscopy using short wavelengths and leave-one-cow-out cross-validation for quantification of somatic cells in milk. *J. Near Infrared Spectrosc.* 2009, *17*, 345–351.
17. Kalinin, A.V.; Krasheninnikov, V.N.; Krivtsun, V.M. Short wave near infrared spectrometry of back scattering and transmission of light by milk for multi-component analysis. *J. Near Infrared Spectrosc.* 2013, *21*, 35–41.
18. APHA-AWWA-WEF. *Standard Methods for the Examination of Water and Wastewater*, 22nd ed.; American Public Health Association (APHA)-American Water Works Association (AWWA)-Water Environment Federation (WEF): Washington, DC, USA, 2012; p. 1496.
19. Kumar, A.; Dhall, P.; Kumar, R. Redefining BOD:COD ratio of pulp mill industrial wastewaters in BOD analysis by formulating a specific microbial seed. *Int. Biodeterior. Biodegrad.* 2010, *64*, 197–202.
20. Savitzky, A.; Golay, M.J.E. Smoothing and differentiation of data by simplified least squares procedures. *Anal. Chem.* 1964, *36*, 1627–1639.

21. Mevik, B.-H.; Wehrens, R. The pls package: Principal component and partial least squares regression in R. *J. Stat. Softw.* 2007, *18*, 1–23.
22. Wold, S.; Sjostrom, M.; Eriksson, L. PLS-regression: A basic tool of chemometrics. *Chemom. Intell. Lab. Syst.* 2001, *58*, 109–130.
23. Curran, P.J. *Principles of Remote Sensing*; Longman: London, UK, 1985; p. 282.
24. R Core Team. *R: A Language and Environment for Statistical Computing*; R Foundation for Statistical Computing: Vienna, Austria, 2013. Available online: http://www.r-project.org (accessed on 8 October 2013).
25. Martens, H.; Naes, T. *Multivariate Calibration*; John Wiley and Sons: Chichester, UK, 1989; p. 419.
26. Willmott, C.J. Some comments on the evaluation of model performance. *Bull. Am. Meteorol. Soc.* 1982, *63*, 1309–1313.
27. Williams, P.C. Implementation of near-infrared technology. In *Near-Infrared Technology in the Agricultural and Food Industries*, 2nd ed.; Williams, P.C., Norris, K., Eds.; American Association of Cereal Chemists: Saint Paul, MN, USA, 1987; pp. 143–167.
28. Pons, M.-N.; Le Bonté, S.; Potier, O. Spectral analysis and fingerprinting for biomedia characterisation. *J. Biotechnol.* 2004, *113*, 211–230.
29. Páscoa, R.; Lopes, J.; Lima, J. *In situ* near infrared monitoring of activated dairy sludge wastewater treatment processes. *J. Near Infrared Spectrosc.* 2008, *16*, 409–419.
30. Yeniay, O.; Göktas, A. A comparison of partial least squares regression with other prediction methods. *Hacet. J. Math. Stat.* 2002, *31*, 99–111.

Chapter 9

Soil Contamination, Risk Assessment and Remediation

Muhammad Aqeel Ashraf[1], Mohd. Jamil Maah[2] and Ismail Yusoff[1]

[1] Department of Geology, Faculty of Science, University of Malaya, Kuala Lumpur, Malaysia

[2] Department of Chemistry, Faculty of Science, University of Malaya, Kuala Lumpur, Malaysia

INTRODUCTION

Environment pollution is a burning topic of the day. Air, water and soil are being polluted alike. Soil being a "universal sink" bears the greatest burden of environmental pollution. It is getting polluted in a number of ways. There is urgency in controlling the soil pollution in order to preserve the soil fertility and increase the productivity. Pollution may be defined as an undesirable change in the physical, chemical and biological characteristics of air, water and soil which affect human life, lives of other useful living plants and animals, industrial progress, living conditions and cultural assets. A pollutant is something which adversely interfere with health, comfort, property or environment of the people. Generally most pollutants are introduced in the environment by sewage, waste, accidental discharge or else they are by-products or residues from the production of something useful. Due to this our precious natural resources like air, water and soil are getting polluted.

The basis of agriculture is Soil. All crops for human food and animal feed depend upon it. We are losing this important natural resource by the accelerated erosion 10 some extent. In addition to this the enormous quantities of man-made waste products, sludge and other product" from new waste treatment plants even polluted water are also causing or leading to soil pollution. In order to preserve the fertility and the productivity of the soil, control measures are to be taken in a herculean manner, thereby improving the health of all living beings.

Assessing the ecological risk of contaminated soil, pesticide application, sewage sludge amendment, and other human activities leading to exposure of the terrestrial environment to hazardous substances is a complicated task with numerous associated problems. Not only is terrestrial ecological risk assessment a relatively new field of science that has developed rapidly only since the mid-1980s, but it is also complicated by the fact that soil, in contrast to most aquatic environments, is very often on private lands and traded as real estate. Professional and economic divergence between the interests of scientists, stakeholders, authorities, engineers, managers, lawyers, nongovernment organizations (NGOs) and regulators is therefore not unusual. Even neglecting those aspects, a number of unresolved problems exist in the way we currently assess risk and manage the impact of anthropogenic substances in the terrestrial environment.

This chapter does not intend to present a comprehensive review of all published data from ecological studies at contaminated sites. Instead, the observations from all case studies are used in the discussion and form the basis for the final conclusion. In each case, we try to answer the following questions:

1. What is soil pollution and how it occurs?
2. How to determine the ecological risk assessment of the soil?
3. To what extent do soil screening levels (over)estimate risk?
4. Do bioassays represent a more realistic risk estimate?
5. Is it possible to make sound field surveys, or do we lack suitable reference situations?
6. What are the possible soil management methods for the polluted soils?

SOIL POLLUTION

Soil pollution is defined as the build-up in soils of persistent toxic compounds, chemicals, salts, radioactive materials, or disease causing agents, which have adverse effects on plant growth and animal health [1].

Soil is the thin layer of organic and inorganic materials that covers the Earth's rocky surface. The organic portion, which is derived from the decayed remains of plants and animal, is concentrated in the dark uppermost topsoil. The inorganic portion made up of rock fragments, was formed over thousands of years by physical and chemical weathering of bedrock. Productive soils are necessary for agriculture to supply the world with sufficient food [2].

There are many different ways that soil can become polluted, such as:

- Seepage from a landfill
- Discharge of industrial waste into the soil
- Percolation of contaminated water into the soil
- Rupture of underground storage tanks
- Excess application of pesticides, herbicides or fertilizer
- Solid waste seepage

The most common chemicals involved in causing soil pollution are:

- Petroleum hydrocarbons
- Heavy metals

- Pesticides
- Solvents

Inorganic Toxic Compounds

Inorganic residues in industrial waste cause serious problems as regards their disposal. They contain metals which have high potential for toxicity. Industrial activity also emits large amounts of arsenic fluorides and sulphur dioxide (SO_2) [3]. Fluorides are found in the atmosphere from superphosphate, phosphoric acid, aluminium, steel and ceramic industries. Sulphur dioxide emitted by factories and thermal plants may make soils very acidic. These metals cause leaf injury and destroy vegetation.

Copper, mercury, cadmium, lead, nickel, arsenic are the elements which can accumulate in the soil, if they get entry either through sewage, industrial waste or mine washings. Some of the fungicides containing copper and mercury also add to soil pollution. Smokes from automobiles contain lead which gets adsorbed by soil particles and is toxic to plants. The toxicity can be minimized by building up soil organic matter, adding lime to soils and keeping the soil alkaline [4].

Organic Wastes

Organic wastes of various types cause pollution hazards. Domestic garbage, municipal sewage and industrial wastes when left in heaps or improperly disposed seriously affect health of human beings, plants and animals [5-7]. Organic wastes contain borates, phosphates, detergents in large amounts. If untreated they will affect the vegetative growth of plants. The main organic contaminants are phenols and coal.

Asbestos, combustible materials, gases like methane, carbon dioxide, hydrogen sulphide, carbon monoxide, sulphur dioxide, petrol are also contaminants. The radioactive materials like uranium, thorium, strontium etc. also cause dangerous soil pollution. Fallout of strontium mostly remains on the soil and is concentrated in the sediments [8]. Decontamination procedures may include continuous cropping and use of chelate amendments. Other liquids wastes like sewage, sewage sludge, etc. are also important sources of soil problems.

Sewage and sewage sludge

Soil pollution is often caused by the uncontrolled disposal of sewage and other liquid wastes resulting from domestic uses of water, industrial wastes containing a variety of pollutants, agricultural effluents from animal husbandry and drainage of irrigation water and urban runoff [9-10]. Irrigation with sewage water causes profound changes in the irrigated soils. Amongst various changes that are brought about in the soil as an outlet of sewage irrigation include physical changes like leaching, changes in humus content, and porosity etc., chemical changes like soil reaction, base exchange status, salinity, quantity and availability of nutrients like nitrogen, potash, phosphorus, etc. Sewage sludges pollute the soil by accumulating the metals like lead, nickel, zinc, cadmium, etc. This may lead to the phytoxicity of plants.

Heavy metal pollutants

Heavy metals are elements having a density greater than five in their elemental form. They mostly find specific absorption sites in the soil where they are retained very strongly either on the inorganic or organic colloids. They are widely distributed in the environment, soils, plants, animals and in their tissues. These are essential for plants and animals in trace amounts. Mainly urban and industrial aerosols, combustion of fuels, liquid and solid from animals and human beings, mining wastes, industrial and agricultural chemicals etc. are contributing heavy metal pollution. Heavy metals are present in all uncontaminated soils as the result of weathering from their parent materials. Concentration of heavy metals in soils and plants is given in Table 1.

Sl.No	**Heavy metal**	**Hithosphere**	**Soil range**	**Plants**
1	Cadmium (Cd)	0.2	0.01-0.7	0.2-0.8
2	Cobalt (Co)	40	1-40	0.05-0.5
3	Chromium (Cr)	200	5-3000	0.2-1.0
4	Copper (Cu)	70	2-100	4-15
5	Iron (Fe)	50,000	7000-5,50,000	140
6	Mercury (Hg)	0.5	0.01-0.3	0.015

7	Manganese (Mn)	1000	100-4000	15-100
8	Molybdenum (Mo)	2.3	0.2-5	1-10
9	Nickel (Ni)	100	10-1000	1
10	Lead (Pb)	16	2-200	0.1-10
11	Tin (Sn)	40	2-100	0.3
12	Zinc (Zn)	80	10-300	8-100

Table 1: Heavy metal concentration in the hithosphere, soils and plants (Ug/gm dry matter)

In agricultural soils, however, the concentration of one or more of these elements may be significantly increased in several ways, like through applications of chemicals, sewage sludge, farm slurries, etc. Increased doses of fertilizers, pesticides or agricultural chemicals, over a period, add heavy metals to soils which may contaminate them. Certain phosphatic fertilizers frequently contain trace amounts of cadmium which may accumulate in these soils. Likewise, some fertilizers when applied to soils, they add certain heavy metals which are given in Table 2.

Table 2: Heavy metal content of fertilizers (ug/gm)

Sl. No	**Fertilizer**	**Co**	**Cr**	**Cu**	**Mn**	**Mo**	**Ni**	**Pb**	**Zn**
1	Nitrochalk	-	-	22	24	-	2	-	15
2	Calcium	0.1	Traces	Traces	Traces	-	-	-	1
3	Nitrate	-	-	To 10	To 5	-	-	-	-
4	Ammonium sulphate	<5	<5	0.800	0.80	<0.05 to 0.22	<5	Traces to 200	0.800
5	Super phosphate	0.02-13	0-1000	Traces to 1000	Traces to 2842	Traces to 35	Traces to 32	Traces to 92	70-3000
6	Potassium chloride	001	-	0-10	Traces-8	<0.05	<1	<1	0-3
7	Potassium sulphate	<5	<5	0-300 to 80	Traces to .33	0.09	<5	<50	<50

Vitamin C and carotene content in vegetables and fruits. The vegetables and fruits grown on over fertilized soil are more prone to attacks by insects and disease.

Indiscriminate use of Pesticides, Insecticides and Herbicides

Plants on which we depend for food are under attack from insects, fungi, bacteria, viruses, rodents and other animals, and must compete with weeds for nutrients. To kill unwanted populations living in or on their crops, farmers use pesticides. The first widespread insecticide use began at the end of World War II and included DDT (dichlorodiphenyltrichloroethane) and gammaxene. Insects soon became resistant to DDT and as the chemical did not decompose readily, it persisted in the environment. Since it was soluble in fat rather than water, it biomagnified up the food chain and disrupted calcium metabolism in birds, causing eggshells to be thin and fragile. As a result, large birds of prey such as the brown pelican, ospreys, falcons and eagles became endangered. DDT has been now been banned in most western countries. Ironically many of them including USA, still produce DDT for export to other developing nations whose needs outweigh the problems caused by it [16].

Dumping of solid Wastes

In general, solid waste includes garbage, domestic refuse and discarded solid materials such as those from commercial, industrial and agricultural operations. They contain increasing amounts of paper, cardboards, plastics, glass, old construction material, packaging material and toxic or otherwise hazardous substances. Since a significant amount of urban solid waste tends to be paper and food waste, the majority is recyclable or biodegradable in landfills. Similarly, most agricultural waste is recycled and mining waste is left on site. The portion of solid waste that is hazardous such as oils, battery metals, heavy metals from smelting industries and organic solvents are the ones we have to pay particular attention to. These can in the long run, get deposited to the soils of the surrounding area and pollute them by altering their chemical and biological properties [17].

Deforestation

Soil Erosion occurs when the weathered soil particles are dislodged and carried away by wind or water. Deforestation, agricultural development, temperature extremes, precipitation including acid rain, and human activities contribute to this erosion. Humans speed up this process by construction, mining, cutting of timber, over cropping and overgrazing. It results in floods and cause soil erosion. Forests and grasslands are an excellent binding material that keeps the soil intact and healthy. They support many habitats and ecosystems, which provide innumerable feeding pathways or food chains to all species. Their loss would threaten food chains and the survival of many species. During the past few years quite a lot of vast green land has been converted into deserts. The precious rain forest habitats of South America, tropical Asia and Africa are coming under pressure of population growth and development (especially timber, construction and agriculture). Many scientists believe that a wealth of medicinal substances including a cure for cancer and aids, lie in these forests. Deforestation is slowly destroying the most productive flora and fauna areas in the world, which also form vast tracts of a very valuable sink for CO_2 [18].

Pollution due to Urbanization

Pollution of surface soils materials (like vegetables, animal wastes, papers, wooden pieces, carcasses, plant twigs, leaves, cloth wastes as well as sweepings) and many non-biodegradable materials (such as plastic bags, plastic bottles, plastic wastes, glass bottles, glass pieces, stone / cement pieces) [19-20]. On a rough estimate Indian cities are producing solid city wastes to the tune of 50,000 - 80,000 metric tons every day. If left uncollected and decomposed, they are a cause of several problems such as;

- Clogging of drains: Causing serious drainage problems including the burst / leakage of drainage lines leading to health problems.
- Barrier to movement of water: Solid wastes have seriously damaged the normal movement of water thus creating problem of inundation, damage to foundation of buildings as well as public health hazards.
- Foul smell: Generated by dumping the wastes at a place.

- Increased microbial activities: Microbial decomposition of organic wastes generate large quantities of methane besides many chemicals to pollute the soil and water flowing on its surface
- When such solid wastes are hospital wastes they create many health problems: As they may have dangerous pathogen within them besides dangerous medicines, injections.

Pollution of underground soil

Underground soil in cities is likely to be polluted by

- Chemicals released by industrial wastes and industrial wastes
- Decomposed and partially decomposed materials of sanitary wastes

Many dangerous chemicals like cadmium, chromium, lead, arsenic, selenium products are likely to be deposited in underground soil. Similarly underground soils polluted by sanitary wastes generate many harmful chemicals. These can damage the normal activities and ecological balance in the underground soil

Effects of Soil Pollution

Agricultural

- Reduced soil fertility
- Reduced nitrogen fixation
- Increased erodibility
- Larger loss of soil and nutrients
- Deposition of silt in tanks and reservoirs
- Reduced crop yield
- Imbalance in soil fauna and flora

Industrial

- Dangerous chemicals entering underground water
- Ecological imbalance

- Release of pollutant gases
- Release of radioactive rays causing health problems
- Increased salinity
- Reduced vegetation

Urban

- Clogging of drains
- Inundation of areas
- Public health problems
- Pollution of drinking water sources
- Foul smell and release of gases
- Waste management problems

Environmental

- Soil becomes unavailable to grow food
- If contaminated soil is used to grow food, the land will usually produce lower yields
- Can cause even more harm because a lack of plants on the soil will cause more erosion
- The pollutants will change the makeup of the soil and the types of microorganisms that will live in it.
- Thus it's possible for soil pollution to change whole ecosystems

Control of Soil Pollution

The following steps have been suggested to control soil pollution. To help prevent soil erosion, we can limit construction in sensitive area. In general we would need less fertilizer and fewer pesticides if we could all adopt the three R's: Reduce, Reuse, and Recycle. This would give us less solid waste.

Extraction and Separation Techniques

In solvent extraction, the contaminated & oil is mixed with an extracting agent in general (an aqueous solution but preferably an organic solvent). Potential applications include the removal of metals such as cadmium, copper, zinc, nickel, chromium, arsenic, antimony and lead using a mineral solution, zinc lead, organo-metallic compounds and some cyanides using sodium hydroxide solution. Hydrocarbons and halogenated hydrocarbons can also be removed [21]. Contamination is often preferentially present in the finer or coarser fraction of the soil or the organic components (ex. humus), contamination can therefore, be removed in some cases using a process which separates the soil into fractions on the basis of specific gravity or particle size or settling velocity.

Thermal Methods

In thermal methods, there are two ways of heat treatment; removal of contaminants by evaporation either by direct heat transfer from heated air or an open flame or by indirect heat transfer, and destruction of the contaminants directly or indirectly at an appropriate temperature. The gas leaving the heating appliance must be treated to destroy or remove any contaminants or unwanted products of combustion. A related process is stream stripping in which steam is injected into soil to aid evaporation of relatively volatile contaminants which may be water soluble or insoluble.

Chemical Methods

Treatment of the soil in suspension in a suitable liquid and without sludging is the two possible methods. In these, intimate, contact between soil and chemical is essential and should be frequently done so that the process of detoxification is complete.

Microbial Treatment Methods

The microbial treatment methods appear to be more promising which can deal with whole range of organic contaminants including phenol, polychlorinated hydrocarbons, oil and oil products, dioxins, etc. There are two different ways of approaching the problems.

- A community of microbes already existing on the site is collected and cultured in the laboratory.

- Strains of microbes are developed in the laboratory that is capable of metabolizing particular chemicals.

Excavation of the soil prior to treatment offers the greatest scope for creating optimum conditions. The excavated soil can be placed on thin layers to various depth using standard earth moving techniques and microbes and nutrients applied using standard agricultural techniques such as fertilizing, ploughing, harrowing, etc.

Reducing Chemical Fertilizer and Pesticide use

Applying bio-fertilizers and manures can reduce chemical fertilizer and pesticide use. Biological methods of pest control can also reduce the use of pesticides and thereby minimize soil pollution.

Reusing of Materials

Materials such as glass containers, plastic bags, paper, cloth etc. can be reused at domestic levels rather than being disposed, reducing solid waste pollution.

Recycling and Recovery of Materials

This is a reasonable solution for reducing soil pollution. Materials such as paper, some kinds of plastics and glass can and are being recycled. This decreases the volume of refuse and helps in the conservation of natural resources. For example, recovery of one tonne of paper can save 17 trees.

Reforesting

Control of land loss and soil erosion can be attempted through restoring forest and grass cover to check wastelands, soil erosion and floods. Crop rotation or mixed cropping can improve the fertility of the land.

Risk-based criteria or standards, developed in the framework of risk oriented policies, are applied to risks estimated with deterministic methodologies, following the steps of hazard characterization, appraisal of exposure and risk characterization, while using exposure-risk relations established beforehand. Risk-based criteria have been applied to decisions about soil remediation in the form of soil clean-up standards [26], to the use of soils for specific purposes and in the United States also to sediment management [27]. The risk-oriented policies considered here [28], assume that background exposure to pollutants carries no risk and that a specified level of soil pollution carries a maximum tolerable or maximum acceptable risk for organisms living locally. The latter is the main basis for standard setting.

In part, risk-oriented soil pollution legislation includes policy goals that are qualitative [29]. For instance, the primary UK legislation on contaminated soil defines land as contaminated in need of risk management 'if significant harm is being caused or there is a significant possibility of such harm being caused' [29]. Mostly, however policies have resulted in specific quantitative values for maximum tolerable or acceptable soil pollution. The analysis of such values used in different industrialized countries has shown that there are very large differences, roughly up to a factor [30]. According to Provoost et al. [31], these differences to a large extent originate in different political choices (e.g. including or excluding ecotoxicity) and in different assumptions as to the modeling of exposure to soil pollutants, including site related factors, such as soil type and building constructions [31].

Risks Related to One Soil Pollutant

In practice, there are several matters which are at variance with the proper establishment of actual risk related to one soil pollutant. These are: the absence of standards for pollutants, neglect of background exposure, and neglect of routes of exposure to soil pollution, neglect of available dose- effect studies and neglect of biological availability. These will now be discussed in more detail.

Absence of Quality Standards

When data regarding soil pollutants are available, they should be compared with quality standards reflecting maximum tolerable risk of exposure. However, such standards are not always in place. For instance, of the volatile organic carbon compounds detected in groundwater samples by the US Geological Service, were unregulated- with no standards in place [32]. Similarly Patterson et al. [33] found a variety of brominated ethenes in Australian groundwater, all lacking standards.

Neglect of Background Exposure

For a proper estimate of soil pollution related risks, exposure to specific soil pollutants should be evaluated in combination with exposure to the same substance that is not related to local soil contamination. Several countries, such as Canada, Germany, Spain and Belgium, do indeed establish soil clean-up standards while considering background dietary and inhalatory exposure but others, e.g. Sweden, Norway and the Netherlands, do not [34]. Neglecting background exposure or specific types of background exposure may have implications for risk estimates.

Neglect of Routes of Exposure to Soil Pollution

In evaluating exposure to soil pollutants, assumptions regarding exposure routes are important. In this respect difference between countries may be noted. Soil clean-up standards for lead of Norway and Sweden differ in part because in Sweden the dominant exposure route is assumed to be by drinking water and in Norway it is thought to be by drinking water and ingestion of soil [34].

Inhalation of household dust and soil particles is not always taken into account in governmental decision making about risks of soil pollution. For instance, in the Netherlands inhalation of soil particles has been neglected as an exposure route, but in e.g. Spain it is not [34]. Neglect of inhalation would seem at variance with existing studies. Nawrot et al. [35] have studied the effects of cadmium pollution in soil (around former thermal zinc plants) and found a significant increase in lung cancer risk correlated with cadmium exposure. They plausibly explain this in terms of exposure of lung tissue to cadmium present in inhaled soil and household dust particles.

Household dust particles have also been found to be important in the exposure of children to pesticides in agricultural settings.

Neglect of Available Dose-Effect Studies

Akesson et al. [37] have analyzed the effects of low environmental cadmium exposure in an epidemiological study of Swedish women in the Lund area, being 53-64 years of age, excluding women from areas with soils heavily polluted by cadmium. Akesson et al. [37] found associations between the internal dose of cadmium and tubular and glomerular kidney effects, which may represent early signs of adverse effects. Women with diabetes seemed to be at increased risk of experiencing such early signs. In view of these data it seems plausible that at a background exposure that is common in Sweden, old women in the general population may be at risk for adverse cadmium effects [34] and that even a modest increase in cadmium exposure due to polluted soil may lead to added risk. However, when establishing soil clean-up standards in Sweden this background exposure has been neglected [34]. Nawrot et al. [35] have studied the relation between mortality and cadmium body burden in Belgium. They obtained evidence that total mortality and non-cardiovascular mortality may be elevated at cadmium body burdens which can be found among the population not living on soils that are currently considered to be a health risk. Similarly there are now strong indications that the negative effects of lead on the neurophysiologic and sexual development may well be found at the level of background exposure common in Western European and US cities [37], though soil pollution policy, at least in European counties, assumes that such background exposure is safe [14].

Ecotoxicological Risks

Maximum acceptable or maximum tolerable ecotoxicological risks are usually derived from a limited number of studies concerning single species under laboratory conditions. Laboratory conditions may be very different from actual conditions in the field, and thus findings in the field are often at variance with laboratory studies [35]. In field studies it has been found that several factors which tend to be neglected in laboratory studies may strongly impact toxic effects of soil pollutants. These include among others: density and adaptability of populations of affected organisms, the presence of other environmental stress factors and the presence or absence of specific landscape elements such as buffer strips [35].

Biological Availability

Biologically available pollutants determine risk [3]. Biological availability may vary strongly for different types of organisms [36]. Biological availability of a compound in a specific soil is also dependent on physical, chemical and biological and spatial factors [35]. Examples of such factors are pH, the amount and nature of organic and mineral compounds also present and the presence of organisms that can mobilize soil pollutants [37-39]. In practice, biological availability may be much at variance with total concentrations [40].

Combination Effects

Limited Accounting of Combination Effects

As to the overall risk of soil pollutants, cumulative effects of the combination of substances present in soils should be considered. However actual standard setting practice has largely focused on criteria relating to one element or compound. In some cases there are criteria for groups of compounds [34]. Such criteria limit the amount (in g/kg soil) of groups of compounds but often do not address the possibility that the risk per unit of weight may be different for different compounds. An exception to this is criteria for the presence of halogenated dioxins and benzofurans and planar biphenyls. The establishment of risk in case of exposure to these compounds uses addition on the basis of equivalent toxicity [40]. This is a major improvement, though it has been pointed out that this approach may still underestimate the risk of neurodevelopment effects [41].

Importance of Combination Effects

Combination effects may be important in two respects. Firstly, coexisting soil contaminants may impact each other's biological availability [43]. Secondly, exposure to a combination of pollutants may be associated with antagonistic, synergistic and additive interactions of these pollutants, impacting their effect on organisms [44-47]. Some risks of pollutant mixtures can be predicted on the basis of existing knowledge. For instance there is a fair chance that there will be dose additivity when effects are receptor mediated [48]. Also in case of narcotic effects, joint-mixture ecotoxicological effects may be predicted [48]. If responses are dissimilar, response addition may be used [49]. A methodology to deal with the ecotoxicity of mixtures giving rise to both dose-additive and response-additive effects has been proposed [48]. This two step model evaluates mixture toxicity for the same mode of action with concentration additivity and the toxicity for different modes of action with response additivity. For determining the severity of ecotoxicological effects in case of heavily polluted soils (in which legal maximum tolerable levels for one or more substances are exceeded), a systematic approach to combination effects based on a mixture of concentration addition and response addition has been proposed [41].

Remedies for Shortcomings

Remedies would seem possible which would allow for a significant improvement in risk estimates. Unregulated substances can get standards. Standards may be regularly updated on the basis of new dose-effect studies. Risk estimates can include both background exposure and all exposure routes for local soil pollution. Estimates of biological availability can be integrated in risk assessments and improved by better testing of bioavailability or by in-vivo monitoring [51]. The deficiencies in taking account of combination effects in ecotoxicity, discussed in section may be addressed by directly testing of ecotoxicity, when the focus is on ecosystem functioning [52,53]. However it should be noted that small effects on the functioning of ecosystems may have large effects over time [35]. This necessitates large numbers of replicate tests that may well be beyond routine practice [35].

In determining combination effects on human health, direct testing on humans is an 'unethical option'. However biomarker-based monitoring of some aspects of soil pollution relevant to humans may be an option. For instance Roos et al. [54] have applied a biomarker based test to original and remediated soils that were contaminated by a variety of polycyclic aromatic hydrocarbons (PAH). They tested the expression profile of cytochromes P 450 [54]. Xiao et al. [55] have measured genotoxic risk of soil contamination using an in-vitro assay with Salmonella. Though the relation between such biomarker-based data gathered and the in-vivo risks awaits further elucidation, the application of tests based on biomarkers for soil pollution is an interesting option in dealing with combination effects on humans.

Also, estimates of risk may be derived from biomarkers which may be monitored in people exposed to soil pollution. Such biomarkers have emerged from epidemiological studies considering the combined effect of substances. An illustration thereof is the study by Lee et al. [56] which found a graded association of the concentration of blood lead and urinary cadmium concentrations with oxidative stress related markers in the US population. This suggests that oxidative stress may be useful as a biomarker for combination effects. It has furthermore been proposed to evaluate effects of exposure to nitroarenes by measuring haemoglobin adducts [57], and of mixtures of volatile organochlorines by measuring glutathione conjugative metabolites [58]. Bioassays based on aryl hydrocarbon (Ah) receptor mediated mechanisms have been proposed which will allow a better alternative to the measurement of polyhalogenated aromatic hydrocarbons [41]. Another option is to estimate risks to human health by taking into account cumulative combination effects in line with established cause-effect relations and research into the effects of actual combinations. It has been shown that risks of compounds with the same targets and the same modes of action may be estimated on the basis of concentration addition, while including toxicity equivalence factors for the compounds involved [59].

This has been shown to apply to receptor-mediated-and reactive mechanisms of toxicity, provided that no chemical reactions occur between the components of the mixture considered [60]. Currently this approach is applied to halogenated dioxins, benzofurans and planar polybiphenyls, though non-linear interactions are not completely absent in this category of compounds [61], and neurodevelopment effects may be underestimated, as pointed out before [41]. Extension of this approach is possible to e.g. polycyclic aromatics, including heterocyclic polycyclic aromatics [18,62] organophosphates that inhibit the enzyme cholinesterase [44,63], compounds that bind to estrogen receptors [64-66], carcinogens [67], a variety of petroleum products [68] and compounds that inhibit the MXR efflux pump [69].

ECOLOGICAL RISK ASSESSMENT

Ecological risk assessment (ERA) is a process of collecting, organizing, and analyzing environmental data to estimate the risk or probability of undesired effects on organisms, populations, or ecosystems caused by various stressors associated with human activities. The basic principles of ecological risk assessment are described in numerous papers [70-72]. All varieties of ERA are associated with uncertainties. The value or usefulness of the different ERA methodologies depends on the uncertainty, predictability, utility, and costs. There are typically two major types of ERA. The first is predictive and is often associated with the authorization and handling of hazardous substances such as pesticides or new and existing chemicals in the European Union. This kind of ERA is ideally done before environmental release. The second type of ERA could be described as an impact assessment rather than a risk assessment, as it is the assessment of changes in populations or ecosystems in sites or areas already polluted. The predictive method is based on more or less generic extrapolations from laboratory or controlled and manipulated semi field studies to real-world situations. The descriptive method is more site specific as it tries to monitor ecosystem changes in historically contaminated soils such as old dumpsites or gas facilities or in field plots after amendment with pesticides or sewage sludge, for example.

Often ERA is performed in phases or tiers, which may include predictive as well as descriptive methods. The successive tiers require, as a rule of thumb, more time, effort, and money. The paradigm or schemes for ERA may vary considerable from country to country, but often consist of an initial problem formulation based on a preliminary site characterization, and a screening assessment, a characterization of exposure, a characterization of effects, and a risk characterization followed by risk management. Although exposure assessment is often just as or even more important, this chapter primarily considers effect assessment. In most European countries, ERA of contaminated soils consists of rather simplified approaches including soil screening levels (SSL) (a.k.a. quality objectives, quality criteria, benchmarks, guideline values) and simple bioassays for a first screening of risk [73-75]. National research or remediation programs have led to the development of a large variety of guideline values.

Although hard to categorize, most fall into two categories: generic or site specific. While the site-specific guidelines require a characterization of pH, organic matter, etc., at the site, generic guideline values are more independent of modifying factors and hence straightforward to legislate. Three major classes of tools for assessing ecological effects may be identified: standardized ecotoxicity experiments with single species exposed under controlled conditions to single chemicals spiked to soil; *ex situ* bioassays, here defined as simple laboratory assays where single species are exposed to historically contaminated soils collected in the field; and finally monitoring, analyzing, and mapping of population or community structures in the field. Furthermore, mesocosm, lysometer, or terrestrial model ecosystems (TME) may be useful; these may be considered as large (multispecies) bioassays or ecotoxicity tests [76-79]. TMEs have the advantage that they operate with the (relatively) undisturbed intrinsic soil populations that make up a small food web. TME hence allow the assessment of effects of toxicants that are mediated through changes in food supply or competition and predation.

One of the keystones in deriving environmental quality criteria is the use of standardized terrestrial test procedures. The emphasis of these prognostic tests is on reproducibility, standardization, international acceptance, and site independence. Although increasing in numbers, relatively few terrestrial tests are still approved by the International Standardisation Organisation (ISO) or Organization for Economic Cooperation & Development (OECD). However, other tests have shown promising results and are likely to be prepared for standardization in the future [80].

However, the major problem in using simple laboratory tests to extrapolate to contaminated land may not be the limitations of test species and the natural variation in species sensitivity. The problems associated with extrapolating from one or a few species, exposed under controlled and typically optimal conditions, to the complex interaction of species and chemicals found in most contaminated ecosystems should also cause concern. Although single-species laboratory tests with spiked materials have their obvious benefits, e.g., they measure direct toxicity of chemicals and interpretation is therefore simple, supplementary tools are often needed. Bioassays, as defined in this context are one of the more frequently used higher-tier alternatives. Basically the same test species may be used in bioassays for assessing the risk of a specific contaminated soil as in standard laboratory tests. However, bioassays have the advantage, compared to the use of spiked soil samples, that the exact toxicity of a specific soil may be accessed directly: this includes the combined and site-specific toxicological effect of the mixture of contaminants and their metabolites. Furthermore, the *in situ* bioavailability of that specific soil is (at least almost) maintained in the laboratory during the exposure period. Several studies have shown a reduction in bioavailability and/or toxicity of soils with an old history of contamination [80-85].

Bioassays are therefore often considered a more realistic tool than generic soil screening levels based on spiked laboratory soils. However, a number of uncertainties or problems may be associated with the use of bioassays and the interpretation of their results. First, the test species are still exposed to the contaminants in a relatively short period compared to the permanent exposure condition found at contaminated sites. Furthermore, they are exposed under more or less optimal conditions, in that stressors such as predation inter- and interspecies competition, drought, frost, and food depletion are eliminated during exposure. Finally, typically only a few species are tested individually.

To compensate for some of the limitations just described, contaminated soil may be assessed using multispecies mesocosms, lysometers, or TME. In these, species interactions may be evaluated by manually introducing several species to the systems or monitoring the intrinsic populations of the soil. Natural climatic conditions may be included if the test system is kept outdoors. However, if we want to get a more realistic and large-scale picture of the impact caused by, for example, pesticide use or sewage sludge application, or to assess the environmental health at waste sites, industrial areas, or gas works, it is often necessary to conduct some kind of field observations. Several case studies exist in which field studies have successfully elucidated the ecological risk of specific activities or the ecological impact at specific sites [85-87]. The small single-species bioassay, large multispecies TME, and field surveys have some drawbacks in common. First of all, it may be difficult to actually link the observed effect to a specific toxic component in the soil. Which of the many substances is actually causing the majority of the observed effects, or is it perhaps a combination of effects? For a hazard classification of soils or a ranking of soils this may not be so important.

However, to evaluate potential risk-reduction measures or risk management procedures it may be important to identify the most problematic substances. A comparison of soil screening values with measured concentrations for each chemical present at a site may be helpful to identify the most likely group of substances causing the observed effect. Other possible tools may include a toxicity identification evaluation (TIE) approach [88]. The TIE approach is a relatively new method, which aims to identify groups of toxicants in soils with mixed pollution. Potentially toxic components present in the soil are fractionated and determined, and the toxicity of each individual fraction is determined by a *Lux* bacteria-based bioassay or the Microtox bioassay. Although perhaps promising, TIE is a time-consuming and hence costly procedure not yet used routinely. Another crucial issue when analyzing the result of bioassays, TME, and field studies is the presence or absence of a proper reference site or soil. The control soil should in principle resemble the contaminated soil in all relevant parameters, e.g., texture, pH, organic matter, waterholding capacity, and nutrient content, a practical problem that very often is difficult to solve. The lack of adequate control or reference sites may, however, be conquered at least partially by the use of multivariate techniques [89], which relate the species composition and abundance to gradients of pollutants. It is not the intention of this chapter to present a review of statistical tools for ecological risk assessment, and hence a detailed discussion about the use of these is not given.

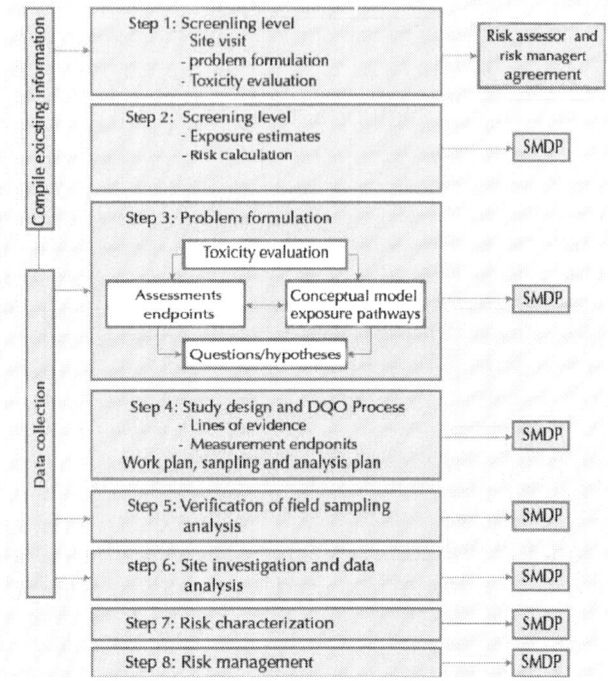

Figure 2: The eight steps in the US-EPA framework for risk assessment of contaminated Superfund sites. DQO = data quality objectives [89].

However, it is obvious that increased computer power and the presence of new easy-to-use software tools have increased the possibility to move away from more conventional univariate statistics such as analysis of variance (ANOVA) to more powerful multivariate statistics that use all collected data to evaluate effects at a higher level of organization. Statistical methods such as the power analysis may also be very useful in planning and designing large-scale ecotoxicity studies such as mesocosms, TME, or field surveys.

The US-EPA has published an Ecological Risk Assessment Guidance, which should be followed when assessing risks at Superfund sites. As all sites are considered unique this should always be done in a site-specific manner. The ERA process suggested by the US-EPA for Superfund sites follows an eight step process, which can be broken down into four categories, i.e. 1) planning and scoping, 2) problem formulation, 3) stressor response and exposure analysis and 4) risk characterisation. Essential for all steps are a negotiation and agreement of the need for further action between the risk assessor, the risk manager and other stakeholders, the so-called scientific-management decision points (SMDP).

SMDP made at the end of the screening-level assessment will not set an initial cleanup goal. Instead, hazard quotients, derived in this step, are used to help determine potential risk. Thus, requiring a cleanup based solely on those values would not be very likely, although it is technically feasible. There are three possible decisions at the SMDP:

- There is enough information to conclude that ecological risks are very low or non-existent, and therefore there is no need to clean up the site on the basis of ecological risk.
- The information is not adequate to make a decision at this point, and the ecological risk assessment process will proceed.
- The information indicates a potential for adverse ecological effects, and a more thorough study is necessary.

In the Netherlands contaminated sites are first determined using a set of soil screening levels called target and intervention values, which take both human and ecological risks into account. At seriously contaminated sites remediation or other soil management decisions are required if the risks cannot be neglected based on a site-specific ecological and human risk assessment, and the chance for dispersion of the contaminants. Until now, the ecological risk assessment has been based on chemical analysis, including a Decision Table harbouring critical dimensions of the impacted area. The United Kingdom and Canada have also developed framework for ecological risk assessment of contamination land. A cornerstone in the UK framework of ERA is the connection to the statutory regime for identification and control of land potentially affected by contamination. The UK framework is based on schemes found in e.g. USA, Canada and the Netherlands. Like these it is a based on a tiered approach where the initial Tier 0 aims to determine whether a site falls under the Part IIA of the legislation. It involves the development of a Conceptual Site Model (CSM), which described what is already (historically) known about the site, e.g. whether there is a likely source-pathway-receptor linkage. The conceptual site model is followed by an initial screening phase (Tier 1) and an actual site-specific characterisation (Tier 2). Tier 1 is a simple deterministic comparison of chemical residue data and the soil quality guideline values supplemented with simple soil-specific toxicity testing.

The final step (Tier 3) involves more detailed *in-situ* studies and for example ecological modelling based on a more advanced ecological theory. Tier 3 is not likely to be conducted at many sites.

Descision Support System for Ecological Risk Assessment

Ecological Risk Assessment is often a complex process with many variables to take into account. ERA involves many stakeholders and all have to be dealt with in a clear and consistent way. A stepwise or tiered approach is therefore useful to overcome the complexity of an ERA. In order to structure all the information collected, a Decision Support System (DSS) can be used. Each tier will lead to a decision to proceed or to stop. A number of decisions supporting systems or frameworks have already been developed in other countries, e.g. UK, the Netherlands and the USA. The DSS presented here is based on basic principles also common in the methodologies used in the USA and UK. However, in the present DSS measures of bioavailability and the use of the Triad approach may be built into the system more systematically. This chapter introduces the overall framework of a novel DSS including the Triad approach and the challenge to weight and scale results used in that process.

Framework for Ecological Risk Assessment

Rutgers et al. (2000) developed a basic flowchart for Ecological Risk Assessment [90], which is used as the backbone of the decision support system (DSS) presented in the Figure 5.

The DSS is separated in three different stages, i.e.

- Stage I. Site characterization and description of land-use.
- Stage II. Determination of ecological aspects.
- Stage III. Site-specific tiered assessment (the Triad):
- – Tier 1. Simple screening
- – Tier 2. Refined screening
- – Tier 3. Detailed assessment
- – Tier 4. Final assessment.

Each of these four tiers is based on a weight of evidence (WoE) approach combining three lines of evidence (Chemistry, (eco) Toxicology and Ecology).

Boundaries of the DSS

The DSS in this chapter is not a full and comprehensive document for managing risk of contaminated land. It focuses strongly on supporting decisions made when considering risk to the terrestrial environment. Therefore it addresses only indirectly the risk to ground water and associated (connected) fresh water systems. Nevertheless information about e.g. reduced bioavailability may be useful when assessing potential risk for leaching of contaminants to ground water or fresh water. Furthermore, it is important to realise that the management of a contaminated site is more than assessing ecological risk. Issues like for example risk for humans, availability and cost of remediation solutions, development plans for the vicinity or the region are equally important.

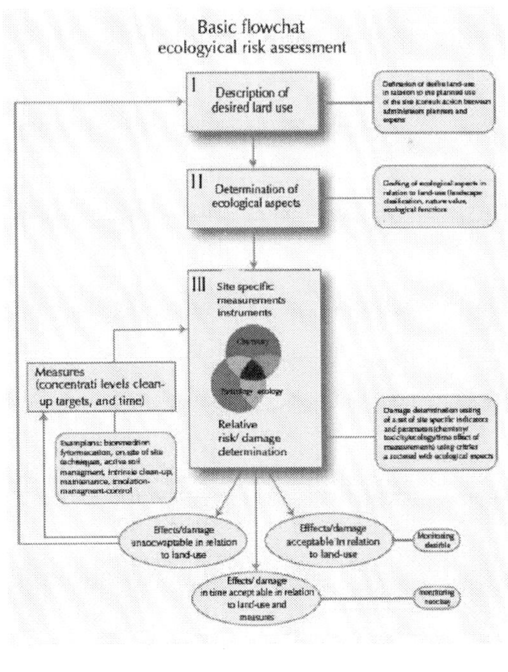

Figure 3: Basic flowchart for ecological risk assessment [90].

Stage I — Site Characterisation and Land-Use Definition

The first step in the DSS is to establish what is often referred to as a Conceptual Site Model. It aims at involving as many stakeholders as possible in order to describe site characteristics and to review all available information from the site, e.g. historical information about land-use, investigation of whether the site may be regulated under specific directives, obvious data gaps and urgency for reaction and data collection. The spatial borders of the site should be defined and the current and the future landuse have to be defined. Consultation between administrators, planners and experts therefore has to take place as early as possible in the process.

Initial Requirements in the DSS

An inquiry among all stakeholders should be conducted as one of the first initiatives. The aim should be to collect as much information about soil characteristics as possible.

Defining land-use

One of the first actions to be taken among all stakeholders is to decide which landuse is required for the site, as this will determine the required data collection and testing. Many land-uses may be defined, but generally the four following overall categories of land-use classes are used:

- industrial area (including infrastructure and pavement).
- urban/residential area (including recreational and green areas).
- agricultural area.
- nature area.

When is an Ecological risk Assessment needed?

Most often a site specific ERA will be initiated only when soil concentrations exceed soil screening levels. However, this may not in itself be a sufficient criterion to go through the entire ERA procedure. Some boundary conditions, based on the present and future type of land-use, the level of contamination and various ecological considerations have to be met in order to rationalize an ERA. The experts and the rest of the stakeholders should answer a number of simple questions in order to conclude whether the required boundary conditions are fulfilled.

Stage II — Determination of Ecological Aspects

At stage II, site-specific ecological features and receptors relating to the land-use defined in Stage I need to be outlined. This includes aspects like key species and life support functions. The potential ecological receptors should be identified in order to determine whether potential source-pathway-receptor linkages can be established. This includes not only ecological receptors directly linked to the site but also those linked indirectly e.g. through leaching of contaminants to connected fresh water systems or (migrating) birds or mammals feeding in the area. In Table 9 some examples are given of land-use and related ecological aspects. This table can be used as a starting point for the selection of ecological aspect. Experts from ecotoxicology and ecology should be involved in the selection of ecological aspects.

Stage III — Site Specific Instruments (The Triad)

If after finalising Stage I and Stage II it is still considered that there is a need for a site specific evaluation of ecological risk the process continues to Stage III using the weight of evidence approach described below.

Weight of Evidence Approaches

In order to deal with conceptual uncertainties in a pragmatic way, it has been proposed to use weight of evidence (WoE) approaches for ERA [90-93]. The rationale is, like in justice, that many independent ways to arrive at one conclusion will provide a stronger evidence for ecological effects, making ERA less uncertain.

In the sediment research area the application of WoE started at an early stage and was called the Sediment Quality Triad. For terrestrial ecosystems WoE approaches and the Triad are still in a developing stage. The Triad approach is based on the simultaneous and integrated deployment of site-specific chemical, toxicological and ecological information in the risk assessment as given in Figure 6. The major assumption is that WoE in three independent disciplines will lead to a more precise answer than an approach, which is solely based on, for example, the concentrations of pollutants at the site. A multidisciplinary approach will help to minimise the number of false positive and false negative conclusions in ERA. It also gives acknowledgement to the fact that ecosystems are too complex to analyse in one-factorial approaches.

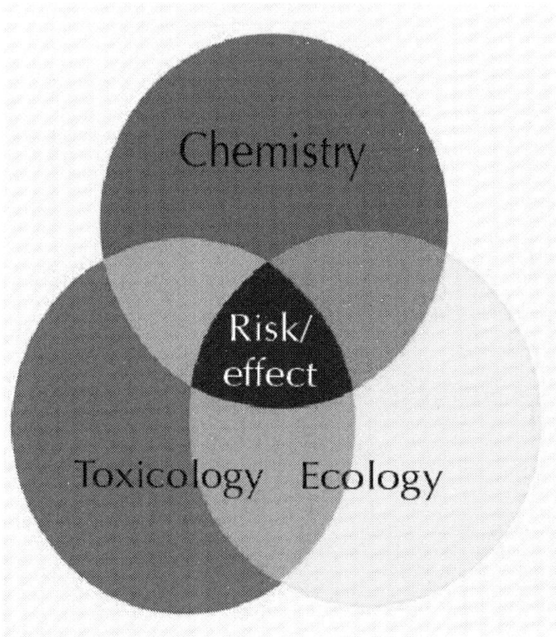

Figure 4: Schematic presentation of the integration of three fields of research according to a Triad [92].

- Chemistry: The concentration of contaminants in the environment (totals, bioavailable), accumulated in biota, or modelled via food-chains is used for calculation of risks on the basis of toxicity data from the literature.

- Toxicology: Bioassays with species across genera are carried out in order to measure the actual toxicity present in environmental samples from the site.
- Ecology: Field ecological observations at the contaminated site are compared to the reference site. Deviations from the reference site, which can be plausibly attributed to the contamination levels, are funnelled into the Triad.

Using the Triad in Site Specific Assessment of Contaminated Soil

Triad is a powerful weight of evidence approach originally developed in order to evaluate sediment quality. In the terrestrial compartment less experience is available on the practical use of the Triad. This chapter describes the use of Triad in more detail and gives an insight into some of the important decisions risk assessors have to make when conducting the Triad in practise, e.g. how to scale, weight and integrate the outcome of the various investigations.

The Triad approach exists of three lines of evidence (LoE), the so-called Triad "legs", i.e. chemistry, (eco) toxicology and ecology. The Triad approach includes a tiered system in which each consecutive tier is increasingly fine-tuned to the site-specific situation. In the first tier the research is simple, broad and generic. In later tiers more specific and complex tests and analyses may be used. For each of the LoE in the Triad there are a variety of analyses or tests that can be chosen. Some examples are:

- Chemistry: Measurement of total concentrations, bioavailable concentrations, bioaccumulation, etc.
- Toxicology: Bioassays (in field and/or in lab), biomarkers etc.
- Ecology: Field observations of vegetation, soil fauna, micro-organisms, etc. In Chapter 6, a number of tests or tools that are for suitable for use in each tier are presented for the chemistry, toxicology and ecology LoE.

DECISION CHARTS IN ECOLOGICAL RISK ASSESSMENT OF CONTAMINATED SITES

Flowcharts

This chapter is an attempt to present a decision support system, which can guide risk assessors in their assessment of site-specific ecological risk. A number of site-specific questions need to be answered before a final decision on performing an ecological risk assessment can be made. This chapter introduces a flow chart for ecological risk assessment of contaminated sites. The flowchart is presented as decision trees as shown in Figure 8 together with a more in-depth introduction to the relevant questions that needs to be addressed and answered when performing a site-specific ecological risk assessment.

Decision Making in Era

The assessment of ecological risk is performed stepwise in tiers. Higher tiers represent gradually more and more complex studies, but also more expensive and laborious studies. The full site-specific risk assessment covers four tiers, i.e.

- Simple screening: Tier 1.
- Refined screening: Tier 2.
- Detailed assessment: Tier 3.
- Final assessment: Tier 4.

The main principle in going from a simple screening over a more refined screening to a detailed assessment of the contaminated site is to minimize time and effort. The actual performance of the risk assessment and use of the various tiers may be very site-specific.

Tier 1 — Simple Screening

After deciding in the two first stages of the ERA that ecological concern needs special consideration, the risk assessment starts typically with a simple evaluation at the screening level. This is done in order to minimize costs until new information indicates the need for further assessment and more sophisticated studies. Therefore, the tools used in the first screening need not only to be reasonably quick and easy, but also relatively cheap. The tools for use in Tier I are described in more detail in the toolboxes C1, T1 and E1. On the basis of the results of instruments used in Tier 1 it is decided to either stop further assessment or continue to a higher tier.

Tier 2 — Refined Screening

Tier 2, still considered being at the screening level, aims at refining the measurement of exposure and at the same time to provide further insight into the toxicological and ecological properties of the contaminated soil. Tier 2 deviate from the conservatism normally associated with the use of total concentration in the risk assessment by taking (rough) estimations of bioavailability into consideration in the chemical LoE. A better screening of the toxicological and ecological properties of the soil compensates for the reduced conservatism in the Chemistry LoE of the Triad. The tools for use in Tier 2 are described in more details in the toolboxes C2, T2 and E2. On the basis of the results in Tier 2 a decision should be made to either stop further assessment or continue to a higher Tier.

Tier 3 — Detailed Assessment

The tools in Tier 3 differ from the ones used in Tier 1 and Tier 2 in that they are more laborious, costly and may take longer. On the other hand they are (often) more realistic and/or ecological relevant in order to give a more comprehensive assessment of the ecological risk at the specific site. The stakeholders should beforehand negotiate a minimum set of tests. Is it for example necessary to consider all trophic levels in the toxicological and ecological LoE? Or does the land-use suggest otherwise? Is it necessary (or possible) to estimate the bioavailability of all the substances exceeding their SSL? If not, how are the non-investigated substances dealt with? The tools described for use in Tier 3 are described in more details in the toolboxes C3, T3 and E3. Depending on the results from Tier 3 a decision should be made to either stop further assessment or continue with an even more detailed assessment in Tier 4

Tier 4 — Final Assessment

In Tier 4, the aim of the studies is to answer any remaining questions and to decrease existing uncertainties and this may often require more in-depth research. Tools in Tier 4 can be similar to tools of Tier 3, but more focus has to be on site-specific circumstances. For example bioassays should be done with organisms, which normally occur at the site. Furthermore, it may be more relevant to consider ecological effects outside the contaminated area on e.g. predators or herbivores feeding in the area or effects in adjacent fresh water systems. This Tier requires specialised knowledge and experience with ERA, which implies that costs can be high and only a limited number of people may be able to perform the tests. Generally only on a very limited number of site evaluations will include investigations at this level. If the results of Tier 4 still indicate risk there are basically two possible solutions. Accept the risk and leave the contamination or remove (parts of) the contamination.

Screening Tools

Triad Based Selection of Methods

For each of the three Lines of Evidence (LoE) in the Triad various methods or tools are available. In order to facilitate the selection of appropriate tools in the right context, the tools have been compiled in subclasses or toolboxes. Each of these is a collection of tools considered to be potentially useful in the designated tiers and LoE of the Triad, i.e. chemistry, toxicology and ecology. Furthermore, the tools are arranged according to their complexity, price and practicability or in other words depending on whether they are most useful for screening or detailed assessment, i.e.

- Toolbox C1. Chemistry tools for simple screening.
- Toolbox T1. Toxicology tools for simple screening.
- Toolbox E1. Ecology tools for simple screening.

- Toolbox C2. Chemistry tools for refined screening.
- Toolbox T2. Toxicology tools for refined screening.
- Toolbox E2. Ecology tools for refined screening.
- Toolbox C3. Chemistry tools for detailed assessment.
- Toolbox T3. Toxicology tools for detailed assessment.
- Toolbox E3. Ecology tools for detailed assessment.
- Toolbox IV. Various tools for the final (Tier 4) assessments.

Toolbox C1 — Chemistry Tools for Simple Screening

At the very first stage of the ERA process, total concentrations of all relevant chemicals are individually compared to soil screening levels (SSL) in order to evaluate whether there is a need for a site specific assessment of ecological risk. In the current Stage III of the ERA, this first generic evaluation of risk is followed by a more site-specific screening of risk including information from all three lines of evidence in the Triad. In the Chemistry part of the Triad more site-specific information is collected by:

Refining and targeting the comparison of soil concentrations with soil related benchmarks for site-specific purposes.

Incorporation of the accumulative risk of a mixture of contaminants by calculating the toxic pressure (TP) of a mixture and by doing so generating more site-specific insight to the potential ecological impact of a contaminated site. Each of these steps can be done separately or in combination, e.g. the TP can be calculated using existing SSL or using new developed benchmarks based on either NOEC or EC50 values or site-specific benchmarks can be compared to soil concentrations individually. The approach entirely depends on the strategy taken by the stakeholder group and the availability of data.

Toolbox T1 — Toxicology Tools for Simple Screening

The main objective of the selected toxicity tests or bioassay at Tier 1 should be to screen the soil for presence of toxic compounds. This includes toxic degradation products or compounds, which are not routinely included in various national analytical programs for contaminated sites. This Tier is the first screening level of the ERA and the cost in form of manpower and money should hence be relatively low.

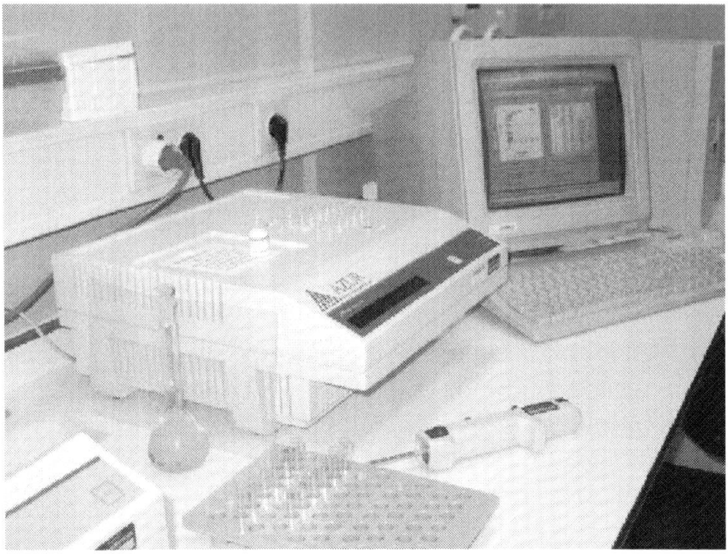

Figure 5: Equipment for measuring luminescence of Vibrio fischeri [94]

Toolbox E1 — Ecology tools for Simple Screening

Ecological surveys or monitoring studies are generally considered a time consuming effort performed by experts. This is in most cases true, wherefore detailed surveys normally take place in higher tier assessment. However, in order to ensure that also ecological information is collected and used in the Triad already in the screening phase, it is recommended to perform a limited examination of the site. A survey of the area with special focus on visible changes in e.g. plant cover or presence or absence of specific plants, trees or scrubs may indicate ecological damage, which can be associated to contaminants present at the site.

Figure 6: Simple survey of the site [95]

If any aerial pictures are available from the area these may give valuable information about the plant cover also historically, which may be helpful in identifying parts of the site where the impact may be highest (hot spots). At this stage the conclusion can in most cases only be indicative. Therefore if the results from the other line of evidence may cause any doubt or the survey indicated potential impact, it is recommended to either continue with a more refined screening in Tier 2 or go directly to the detailed assessment in Tier 3.

Toolbox C2 — Chemistry Tools for Refined Screening

Selective solvent extraction

It may be considered useful to adjust the estimate of exposure by taking bioavailability into consideration and hereby deviating from the conservatism normally associated to the use of total concentration in the risk assessment. The principle in this refinement of the ecological risk assessment is to extract a more ecotoxicologically relevant fraction of the contamination than the total concentration. The latter generally tends to overestimate the risk of historically contaminated soils. In this screening phase no attempt is made to estimate the freely dissolved or readily bioavailable concentration of contaminants. Table 10 explains principal studies that employed chemical extractants to evaluate bioavailability.

Contaminant (Ref)	Solvent	Bioassay	Operation	Comments
Atrazine Phenanthrene (1)	MethanoVwater, n-Butanol	Earthworm uptake	25 ml extractant, 10 g solid.	Methanol/ water best predictor for
		Degradation	Shaking for 2 h	atrazine, n-butanol for phenanthrene.
DDT. DDE, DDD PAH (mixture) (2)	THE Ethanol	Earthworm uptake	15-20 ml extractant, I g soil, 10 sec mixing	Good correlation with earthworm accumulation.
Anthracene, Fluoranthene, Pyrene (3)	n-Butanol Propanol Ethyl acetate	Plant retention Earthworm uptake Degradation	25 ml extractant, 1-2 g soil, 5 sec mixing	Reasonable correlation with bioassays.
Phenanthrene Pyrene Chrysene (4)	n-Butanol	Earthworm uptake Degradation	15 ml extractant, 5-10 g soil, Mixing: 5 sec (worm) or 120 sec (degradation)	Applicable for bioavailability prediction.

Table 5: Outline of principal studies that employed chemical extractants to evaluate bioavailability.

Instead the fraction of the contaminants is extracted, which can be directly compared to the existing soil screening levels. This is considered to be a relatively simple and quick method to screen for potential risk of contaminants in a more realistic way than using total concentrations. The extracted concentration (mg kg-1) is compared to the SSL and the result used in the Triad. It is therefore a prerequisite of this comparison that the extractability in the tests (with spiked soils) used for deriving SSL is close to 100% by the methods used. In most short-term tests (< four weeks) it will be reasonable to assume that only little "true" ageing or strong sequestering occurs and hence a majority of the spiked chemicals are still extractable with mild organic solvents. However, for most methods this still has to be fully validated.

Organic solvents most frequently used include methanol/water in different ratios, nbutanol, ethanol, propanpol, ethyl acetate and tetrahydrofuran (THF) given in Table 10. The method establishes preferential partitioning of hydrophobic contaminants to the extractant by increasing their solubility in the aqueous phase whilst removing pollutant compounds from soil surfaces establishing equilibrium conditions. No standard protocol has been adopted for mild chemical extractions in relation to bioavailability testing. Common methodology in literature primarily includes a soil sample to which a volume of chemical extractant is added (generally 1 – 10 g soil,15 – 25 ml extractant). This is followed by a period for mixing, e.g. vigorous mixing for10 – 120 seconds or shaking by orbital shakers for up to 2 hours. The extraction studies have mostly involved PAH and insecticides (including DDT, DDE, DDD and atrazine) [93-95]. Studies that haverelated extractability with results from bioassays have generally focused on uptake and accumulation (% taken up by earthworms or plants) and bacterial degradation (%removed). Therefore, since convincing relationships between the chemical and biological tests were found it may indicate a potential for such extraction methods to predict bioavailability.

Toolbox T2 — Toxicology Tools for Refined Screening

In the first simple screening of Tier I focus was on marine bacteria and aquatic/sediment living species. In Tier 2 relatively simple tests with soil dwelling species are used for a more refined screening of the soil samples, i.e. the earthworm survival tests and avoidance tests using soil invertebrates.

Figure 7: Organisms screening in soil [98]

The habitat function of soils is often assessed using the reproduction test with *Eisenia fetida*. The avoidance test with *Eisenia fetida* is a suitable screening test, which is less cost-intensive in terms of duration and workload than the reproduction test, and at the same time (normally) more sensitive than the acute test with the same species.

Toolbox E2 — Ecology tools for Refined Screening

In Tier 2 the observations from the survey may be expanded by simple on-site assessment of the overall soil functioning or biological activity of the soils. Recommended tools include bait-lamina sticks and simple microbial tests using general endpoints like soil respiration or C/N mineralization rates.

Figure 8: Bait-lamina sticks [99]

The main principle for tests at this level is to be relatively simple and cheap but at the same time to give valuable information whether or not the soil has lost some of its main services. Bait-lamina sticks for example have been demonstrated useful for describing biological activity of the soils in a general matter.

Toolbox C3 — Chemistry tools for Detailed Assessment

The objective of the tools found in this toolbox is to assess the bioavailable and freely dissolved fraction of pollutants found in pore water of soils from contaminated sites. The methods should (in principle) be able to mimic the fraction of organic pollutants available for uptake in biota. The collection of methods includes various non-depleting and depleting pore water extractions. Very few terrestrial ecotoxicity data are yet expressed as e.g. pore water concentrations. Instead, the outcome of the methodologies in this toolbox is compared with water quality standards.

Toolbox T3 — Toxicology Tools for Detailed Assessment

The objective of the tools found in this toolbox is to evaluate the potential impact of contaminated soils to fauna and plants and hereby the entire ecosystems. Some of the methods use introduced, and not intrinsic, species. The benefit of this is a higher degree of standardisation, as the species used in these bioassays is easy to maintain in laboratory cultures compared to naturally occurring species. The drawback may be that their ecological relevance is less obvious. For example the compost worm *Eisenia fetida* is used as a surrogate to evaluate risk to soil dwelling earthworms. Two sets of bioassays are presented. One for directly assessing potential risk for soil dwelling species, including micro-organisms, plants and soil invertebrates, and one for assessing indirectly risk to aquatic species through e.g. leaching of contaminants. It is often anticipated that soil organisms are exposed to pollutants mainly through uptake from pore water. Therefore it may also be possible to evaluate, or at least to compare or rank, the risk of contaminated soil samples to soil dwelling organisms on the basis of the outcome of the aquatic test using elutriate or pore water. The choice of bioassays depends on a number of variables,

e.g.:

- The current and future land-use, i.e. targets of protection.
- The size of the contaminated area.
- The potential for ground water or surface water contamination.
- The need of many simple tests or fewer more complicated tests.

Figure 9: Simple plant tests [91]

Toolbox E3 - Ecology Tools for Detailed Assessment

In this late tier of the Triad, the objective of the activities is community or population response analysis, typically by conducting field surveys. As these studies (most often) are time consuming, costly and dependent on ecologically, taxonomically and statistical expertise they are most frequently done on large-scale sites with a long-term- remediation perspective. In fresh water ecosystem community surveys have been widely used with relative success. The absence of species from places where they would be expected to occur could be a strong identification of unacceptable levels of contaminants. However, this type of studies has only seldom been used for the terrestrial environment. The reasons for this are many. One of the dominants may be the lack of a concentration gradient and obvious "upstream" reference sites at most contaminated areas. No world-wide accepted guideline on how to plan and perform a terrestrial field survey is available and hence no straight-forward and easy-to-follow description can be given. The decision on when, where and how to conduct field surveys depends on a number of issues, e.g. the size of the area, the land-use, the type of contaminants present, time of the year and last but not least the time and money available to perform the study. Nevertheless, a number of general considerations have to be made in the planning phase of a successful field survey. These include (but are not limited to):

- Identify the targets of concern and the species to monitor.
- Elucidate the natural temporal and spatial variation before initiating a field study.
- Use statistical (power) analyses to determine the minimum number of samples or replicates needed to emonstrate the decided difference, e.g. 25% change.
- In order to establish a cause-effect relationship, a number of confounding parameters need to be characterized both at the reference and the test site, e.g. soil type, pH, salinity, hydrology, nutrient- and organic matter content and the presence of other contaminants.

As no single description on how to perform ecological surveys for contaminated sites can be given, some general considerations and useful references for this tier of the ecological risk assessment are given below for:

- Assessing impact in the overall biological activity and organic matter breakdown.
- Assessing impact on the microbial community.
- Assessing impact on the plant community.
- Assessing impact on the invertebrate community.

Reference Data from Reference sites, Reference Samples and Literature

A crucial factor in a risk assessment is the quality of reference data, because the results of the site-specific ecological measurements or calculations are compared against these data. This is true for as well chemical information (i.e. background levels in that region), toxicological data from bioassays (i.e. site relevant reference soil and well characterized control soil in order to verify the test performance) and ecological field surveys.

Figure 10: Litter bags [94]

The reference soil should in principle resemble the contaminated soil in all relevant parameters, e.g. texture, pH, organic matter, water-holding capacity, nutrient content. In practice, these ideal spots are difficult to find. If there is no or inadequate reference information, effects can only be determined in relative terms by comparison with other sites. This is usually adequate for determining the degree of urgency and/or the need for remediation. Reference data can be obtained by including reference sites (preferably more than one) in the sampling scheme, including reference measurements in the experimental set-up, or by obtaining reference data from the literature or by expert based judgment.

Higher tier assessment of the impact on biological activity and organic matter breakdown

In addition to the general information about biological activity in soils generated in Tier 2 from the bait-lamina test, other, slightly more laborious, tests may give additional information about the overall biological activity in soil, e.g. wheat straw decomposition (litter bag test) and cotton strip degradation. A review paper from Van Gestel et al. (2003) concluded that while the bait-lamina gave the best reflection of the biological activity of soil animals, e.g. earthworms, springtails and enchytraeids, the litter bag test and the cotton strip test are more indicative of the microbial activity in the soil [96]. Knacker et al. (2003) reviewed the usefulness of five different litter decomposition tests and concluded that the litter-bag test had distinct advantages over the others [97]. All of these simple tests only give insight into the overall activity in soils and the breakdown of organic material. They are hence most suitable on their own in cases of land-use with low sensitivity, e.g. industrial land. For land-uses where structural endpoints, e.g. biodiversity or specific species, are the target of protection other endpoint(s) should be monitored as well.

Higher Tier Assessment of the Impact on the Microbial Community

The number of microorganisms, especially bacteria, in soil is extremely large. They differ widely in their function and sensitivity to chemicals. Besides more classical (and simple) measures of the microbial community like total bacterial biomass, the number of colony forming units and substance induced respiration rate (SIR), more advanced methods for assessing the impact of contamination on soil microorganisms have recently been made available.

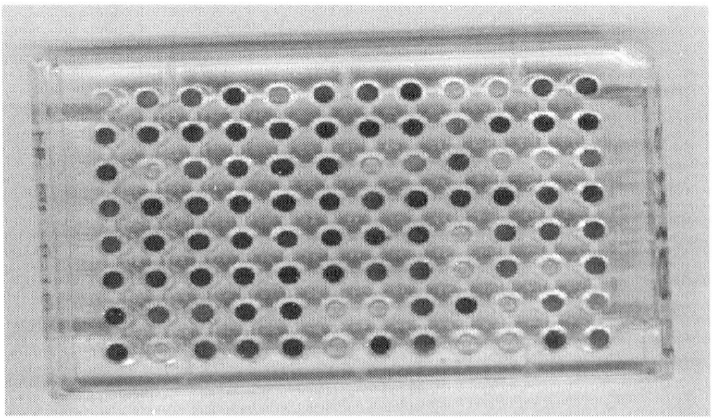

Figure 11: Microbial community [96]

This include microbial "fingerprinting" like phospholipid fatty acid analysis (PLFA) and community-level physiological-profiling (CLPP) based on metabolic response using BIOLOG plate systems, and the use of pollution induced community tolerance (PICT).

Higher tier Assessment of the Impact on the Plant Community

Plants interact dynamically with the physical and chemical characteristics of soils. Soil types and site characteristics, therefore, greatly influence the occurrence of plants and their total aboveground biomass (also called, Net Primary Production, i.e. NPP) within given climatic conditions and human management. Vegetation cover is an important indicator of soil quality and a diverse plant community is normally a good indication of essential soil functions such as the decomposition process, the mineralisation rate, and the occurrence of soil dwelling animals associated to (fresh) organic matter. Vascular plants are easy to sample. They are immobile and hence associated to soil contamination (and airborne pollution). Plant community parameters like plant cover, aboveground plant biomass, plant shoot/ root ratio, species diversity and the binary occurrence (presence/ absence) of specific indicator species like metal-tolerant species may be used successfully in ERA.

Figure 12: Plant survey [95]

Higher Tier Assessment of the Impact on the Soil Invertebrate Community

Survey of soil biota in order to evaluate the effect of various sources of pollution on soil communities on historically contaminated sites have not yet been used on a larger scale by e.g. consultants. However, numerous (monitoring) studies by various research groups can be found in the open literature. Methods of surveying include:

- Collection of soil samples followed by extraction in the laboratory.
- Extraction or collection of organisms in the field, e.g. by hand-sorting or by the application of mustard or formalin.
- Trapping (surface dwelling) animals by the use of e.g. pit-falls. Monitoring species includes earthworms, snails, oribatid mites, nematodes, springtails, ants, ground-living beetles and spiders. Most of the studies have been done on metal contaminated sites (see references below). A substantial amount of work has been put into the challenge of developing a soil invertebrate system for evaluating risk of pollutants. The only soil invertebrate system that is used on a regularly basis in the context of ecological risk assessment of contaminated soils is most likely the nematode Maturity Index (MI). The system is based on the evidence, that rapid colonising species dominate nematode communities in disturbed ecosystems. In the Netherlands experience with surveys of soil invertebrates from the monitoring programme Biological Indicator for Soil Quality (BISQ) has also been used in ERA.

Figure 13: Soil fauna sampling [98]

Toolbox for Tests in Tier IV

The final assessment in the ERA process is not likely to be initiated for many contaminated sites. The choice of additional tests or monitoring at this level of the ERA is bound to be very site-specific and hence an issue for negotiation between stakeholders and experts. Accumulation in biota is included in this toolbox as the internal concentration in biota is believed, at least to some extend, to reflect uptake and then bioavailability. An alternative in this final tier could also be to model uptake in biota provided sufficient data is available [98].

Figure 14: Detailed field survey [93]

SOIL MANAGEMENT PROGRAM (REMEDIATION)

Reporting Norms and Standards for Contaminated Land

The aim is to investigate whether the land has been contaminated, and if contamination has occurred whether the contamination presents a significant risk of harm. The commonly encountered international practice consists of three distinct reporting phases. It progresses from Phase 1 desktop and site walkover assessments with limited investigation and testing to a Phase 2 detailed invasive investigation and testing for site characterisation to a comprehensive Phase 3 report with an evaluation of remediation objectives and a proposed remediation plan, supported by control and monitoring measures for the activities.

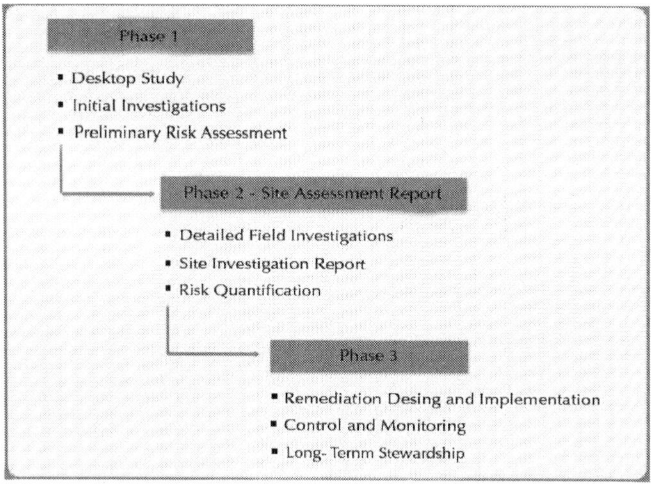

Figure 15: A phase approach for the assessment and remediation of contaminated land

The reporting system requires norms and standards of practice to be strictly applied, but also must retain flexibility to allow for decisions on the contaminated status of sites to be made in the most beneficial manner (considering ecological, social and economic aspects) also taking into account timeframes. In some cases, urgent priority works may require that the phased approach to reporting has to move forward in a concurrent single report.

REQUIREMENTS FOR PRELIMINARY SITE ASSESSMENT PHASE 1

A preliminary site assessment must consider the following elements

Site description-location and size

- Nature and extent of contamination, contaminants of concern and historic activities that may be sources of contamination. List all present and past activities at the site that involves storage and production, use, treatment or disposal of hazardous material that could contaminate the site.
- Describe current condition of the site and contents and the results of any previous assessment report
- • Local topography and geology, drainage, surface cover, vegetation.
- Status of ground water, approximate depth to water table
- Proximity to surface water
- Proximity to drinking water supplies
- Annual rainfall and flood potential
- Land and water use for the nearby areas and site
- Any other regulations as Regulated by the ministry

All data may not be available, or data may vary in terms of uncertainty, it is thus important to recognize gaps in the knowledge base and to decide whether additional data must be obtained on the site characterisation. This may trigger the commencement of Phase 2 Investigations. The Phase 1 report must make clear recommendations on the status of the contamination risk posed by the site. If a complete site history clearly demonstrates that the site activities do not pose a contamination threat then no further investigation is warranted and the site should be recommended as suitable for reuse. In most cases it is likely that some level of preliminary investigation will be required to provide the level of certainty required to enable property re-development or transfer. A limited investigation of certain subsurface activities, for example, underground storage tanks, would be necessary to obtain a waiver on the contamination status of a site at a Phase 1 level of reporting.

Approval Requirements

If soil contaminants are found at concentrations that exceed the applicable standards specified in the approval, the approval holder is required to implement a soil management program. The program must first address source control to stop on-going contaminant releases. After the sources of contamination have been stopped, further assessment and delineation of the contaminated area may be necessary. When the extent of contamination is understood, remediation objectives, as described below, must be adopted for the area. Once remediation objectives have been agreed to, appropriate treatment or containment technologies can be chosen and the management plan finalized.

Developing Remediation Objectives

Remediation objectives may be developed in a variety of ways ranging from generic guidelines to site-specific risk assessment. Generic guidelines are numerical concentration limits that are applicable under a variety of site conditions. When neither Environmental Protection nor the agency has a guideline for a particular substance, four options are available. First, the remediation objective may be based on the ambient background concentration for the site. Second, the guideline development protocol may be applied. Third, a remediation objective may be adopted from another jurisdiction if the proponent can show that the remediation objective is consistent with the environmental protection goals of the approval. Finally, a remediation objective may be developed by the proponent using site-specific risk assessment procedures. A risk assessment for an approved facility will focus on human health concerns but fundamental ecological concerns must also be addressed. Site-specific risk assessment is a means of quantifying the likelihood that soil contamination will have a harmful effect under conditions found at a specific site. The essential components of human health and ecological risk assessments are similar; however, ecological risk assessments tend to be more complex than those for human health because a wide variety of receptors may have to be considered. Very briefly, a site-specific risk assessment may be described as consisting of the following steps or components as shown in Figure 20;

Problem formulation involves developing a conceptual model of the possible contaminant effects on receptors at the site. The conceptual model describes contaminant distribution and concentration in relation to the receptors and their patterns of activity on the site. The *exposure assessment* describes the pathways by which soil contaminants may be taken up by the receptor. This information is combined with receptor characteristics in order to estimate the contaminant uptake. The *toxicity assessment* describes the adverse effects that the contaminants may cause and the dose at which these effects occur.

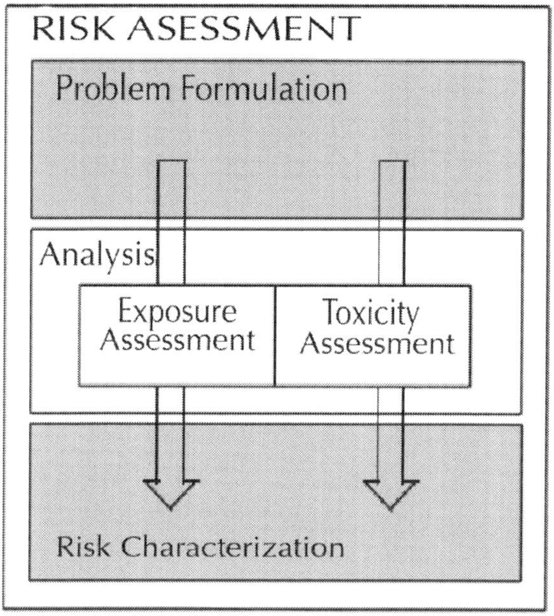

Figure 16: Risk Assessment Framework

The final step, *risk characterization*, compares this dose with the uptake rate (estimated during the exposure assessment) and determines whether or not an adverse effect is likely to occur. Risk assessment procedures can also be used to back calculate a contaminant concentration in soil at which no adverse effects are expected. At times, it may not be possible to remediate to a level compatible with industrial land-use objectives. In such cases, the approval holder must ensure that the contaminants are contained and receptor exposure does not occur. Formal risk assessment procedures are necessary to meet these requirements. Typically, an engineered containment system is required, the performance of which must be confirmed through periodic inspections and monitoring. The approval holder bears responsibility for the design, construction, assessment and maintenance of the risk management system, and any necessary emergency response should the system fail.

Enforcement Policy

Enforcement is generally the last step in a regulatory process that aims to first prevent potential problems before they arise and to resolve existing problems in a cooperative manner. When discussions between the approval holder and the Department fail to prevent or resolve a contravention of an approval or the Act, a number of abatement and enforcement tools are available to the Department including:

- environmental protection orders;
- warning letters;
- tickets;
- administrative penalties;
- enforcement orders;
- prosecutions;
- court orders; and
- cancellation of approvals or certificates.

Due to various factors, including the high cost involved in remediation interventions, it is essential that a holistic and tiered, risk-based approach be adopted that is founded on international best practice, to address remediation in a uniform manner across the country. This is relevant irrespective of the sector of occurrence to safeguard both human health and the natural environment. The Framework is based on a review of international practice in the developed countries of the world and the emergence of remediation policy from developing countries, and an assessment of alternative approaches and methodologies that may find application in the development of a remediation framework.

CONCLUSION

Soil pollution is a result of many activities and experiments done by mankind which end up contaminating the soil. Industrial wastes such as harmful gases and chemicals, agricultural pesticides, fertilizers and insecticides are the most common causes of soil pollution. The others are ignorance towards soil management and related systems, unfavourable and harmful irrigation practices, improper septic system and management and maintenance of the same, leakages from sanitary sewage. There is urgent need for a tiered approach in ecological risk assessment of contaminated soils. Generic soil screening levels are needed as a first tier. Higher tiers of ecological risk assessment should, however, contain some kind of site-specific assessment. It is furthermore important to organize the various studies in a framework or decision support system that is transparent and useful for all stakeholders. A weight of evidence approach may be an obvious choice to deal with these uncertainties. The TRIAD approach, which incorporates and categorizes information in a triangle – chemistry, toxicology, and ecology – is an appropriate tool for handling conceptual uncertainties. Several remedies to these shortcomings have been proposed. Regarding ecotoxicity direct testing would allow for a major improvement in risk estimates. As to human health risks: including biological availability in risk estimates, more use of up to date knowledge about exposure routes, dose-effect relations and combination effects, and biomonitoring of effects are options for improvement.

ACKNOWLEDGEMENTS

Authors greatly acknowledge the University Malaya Research Grant (RG257-13AFR). Moreover, the comments of two anonymous reviewers are gratefully acknowledged.

REFERENCES

1. Okrent D. 1999. On intergenerational equity and its clash with intragenerational equity and on the need for policies to guide the regulation of disposal of wastes and other activities posing very long time risks. Risk Analysis 19: 877-901.
2. Belluck, D.A., Benjamin, S.L., Baveye, P., Sampson, J., Johnson, B. 2003. Widespread arsenic contamination of soils in residential areas and public spaces: an emerging regulatory or medical crisis? International Journal of Toxicology 22: 109-128.
3. Richardson, G.M., Bright, D.A., Dodd, M. 2006. Do current standards of practice in Canada measure what is relevant to human exposure at contaminated sites? II: oral bioaccessibility

of contaminants in soil. Human and Ecological Risk Assessment 12: 606-618.

4. Van Zorge, J. A. 1996. Exposure to mixtures of chemical substances: is there a need for regulations? Food and Chemical Toxicology 34, 1033-1036.
5. Swartjes, F.A. 1999. Risk-based assessment of soil and groundwater quality in the Netherlands: standards and remediation urgency. Risk Analysis 19:1235-1248.
6. Eikelboom, R.T., Ruwiel E., Goumans, J.J.M. 2001. The building materials decree: an example of a Dutch regulation based on the potential impact of materials on the environment. Waste Management 21: 295-302.
7. Crane, M. and Giddings, J.M. 2004. 'Ecologically acceptable concentrations' when assessing the environmental risks of pesticides under European Directive 91 414/EEC. Human and Ecological Risk Assessment 10: 733-747.
8. Nathanail, P., McCaffrey, C., Earl, N., Forster, N.D., Gillett, A.G., Ogden, R. 2005. A deterministic method for deriving site-specific human health assessment criteria for contaminants in soil. Human and Ecological Risk Assessment 11: 389-410.
9. Tarazona, J.V., Fernandez, M.D., Vega, M.M. 2005. Regulation of contaminated soils in Spain. Journal of Soil and Sediments 5:121-124.
10. Evans, J., Wood, G., Miller, A. 2006. The risk assessment-policy gap: An example from the UK contaminated land regime. Environment International 32: 1066-1071.
11. Huinink, J.T.M. 1998. Soil quality requirements of use in urban environments. Soil and Tillage Research 47: 157-162.
12. Urzelai, A., Vega, M., Angulo, E. 2000. Deriving ecological risk-based soil quality values in the Basque Country. Science of the Total Environment 247: 279-284.
13. Apitz, S.E. 2008. Is risk based, sustainable sediment management consistent with European policy. Journal of Soils and Sediments 8: 461-466.
14. Provoost, J., Cornelis, C., Swartjes, F. 2006. Comparison of soil clean-up standards fort race elements between countries: why do they differ? Journal of Soil and Sediments 6: 173-181.

15. Provoost, J., Reijnders, L., Swartjes, F., Bronders, J., Carlon, C., D'Allessandro, M., Cornelis, C. 2008. Parameters causing variation between soil screening values and the effect of harmonization. Journal of Soils and Sediments 8: 298-311.
16. Toccalino, P.L. and Norman, J.E. 2006. Health-based screening levels to evaluate US Geological Survey groundwater quality data. Risk Analysis 26: 1339-1348.
17. Patterson, M.M., Cohen, E., Prommer, H., Thomas, D.G., Rhodes, S., McKinley, A.I. 2007. Origin of mixed brominated ethane groundwater plume: contaminant degradation pathways and reactions. Environmental Science & Technology 41: 1352-138.
18. Leon Paumen, M. 2008. Invertebrate life cycle responses to PAC exposure. PhD thesis. Amsterdam: University of Amsterdam.
19. Nawrot, T., Plusquin, M., Hogervorst, J., Roels, H.A., Celis, H., Thijs, L., Vangronsveld, J., Van Hecke, E., Staessen, J. 2006. Environmental exposure to cadmium and risk of cancer: a prospective population-based study. The Lancet Oncology 7: 119-126.
20. Simcox, N.J., Fenske, R.A., Wolz, S.A., Lee, I.W., Kalman, D.A. 1995. Pesticides in household dust and soil: exposure pathways for children of agricultural families. Environmental Health Perspectives 103: 1126-1134.
21. Laidlaw, M.A.S., Mielke, H.W., Filippelli, G.M., Johnson, D.L., Gonzales, C.R. 2005. Seasonality and children's blood lead levels: developing a predictive model using climate variables and blood data for Indianapolis, Indiana, Syracuse, New York and New Orleans Louisiana (USA). Environmental Health Perspectives 113: 793-800.
22. Laidlaw, M.A.S., Mielke, H.W., Filippelli, G.M., Johnson, D.L. 2006. Blood lead in children: Laidlaw et al. respond. Environmental Health Perspectives 114: A 19.
23. Rothenberg, S.J. and Rothenberg, J.C. 2005. Testing the dose-response specification in epidemiology: public health and policy consequences for lead. Environmental Health Perspectives 113: 1190-1195.
24. Akesson, A., Lundh, T., Vahter, M., Bjellerup, P., Lidfeldt, J., Nerbrand, C., Samisoe, G., Strömberg, U., Skerfving, S. 2005.

Tubular and glomerular kidney effects in Swedish women with low environmental cadmium exposure. Environmental Health Perspectives 113: 1627-1631.

25. Nawrot, T.S., van Hecke, E., Thijs, L., Richart, T., Kuznetsova, T., Jin, Y., Vangronsveld, J., Roels, H.A., Staessen, J.A. 2008. Cadmium-related mortality and long-term secular trends in the cadmium body burden of an environmentally exposed population. Environmental Health Perspectives 116: 1620-1628.
26. Lamphear, R.P., Dietrich, K., Auinger, P., Cox, C. 2000. Cognitive defects associated with blood lead concentrations, 10 microgram/dl in United States children and adolescents. Public Health Reports 115: 521-529.
27. Wang, C.L., Chuang, H.J., Ho, C.K., Vang, C.Y. 2002. Relationship between blood lead concentrations and learning achievement among primary schoolchildren in Taiwan. Environmental Research A 89: 167-181.
28. Wu, T., Buck, G.M., Mendola, P. 2003. Blood levels and sexual maturation in US girls. Environmental Health Perspectives 111: 737-741.
29. Von Storch, H., Costa-Cabral, M., Hagner, C., Freser, F., Pacyna, Y., Pacyna, E., Kolb, S. 2003. Four decades of gasoline lead emissions and control policies in Europe: a retroactive assessment. Science of the Total Environment 311: 151-176.
30. Mielke, H.W., Gonzales, C.R., Powell, E., Jartun, M., Mielke Jr., P.W.2007. Nonlinear association between soil lead and blood lead of children in Metropolitan New Orleans, Louisiana: 2000-2005. Science of the Total Environment 388: 43-53.
31. Provoost, J., Cornelis, C., Swartjes, F. 2006. Comparison of soil clean-up standards for trace elements between countries: why do they differ? Journal of Soils and Sediments 6: 173–181.
32. Miranda, M.L., Kim, D., Overstreet-Galeano, M.A., Paul, C.J., Hull, A.P., Morgan, S.P. 2007. The relationship between early childhood blood lead levels and performance on end-of-grade tests. Environmental Health Perspectives 115: 1242-1247.
33. Patterson, B. M., Cohen, E., Prommer, H., Thomas, D. G., Rhodes, S., McKinley, A. J. 2007. Origin of a mixed brominated ethene groundwater plume: contaminant degradation pathways and reactions. Environmental Science & Technology 41: 1352–1358.

34. Huel, G., Sahuquillo, J., Debotte, G., Oury, J., Takser, L. 2008. Hair mercury negatively correlates with calcium pump activity in human term newborns and their mothers at delivery. Environmental Health Perspectives 116: 263-267.

35. Nawrot, T., Plusquin, M., Hogervorst, J., et al., 2006. Environmental exposure to cadmium and risk of cancer: a prospective population-based study. The Lancet Oncology 7: 119–126.

36. Laidlaw, M. A. S., Mielke, H. W., Filippelli, G. M., Johnson, D. L. 2006. Blood lead in children: Laidlaw et al. respond. Environmental Health Perspectives 114: A19-A23.

37. Akesson, A., Lundh, T., Vahter, M., et al., 2005. Tubular and glomerular kidney effects in Swedish women with low environmental cadmium exposure. Environmental Health Perspectives 113: 1627–1631.

38. Mergler, D., Anderson, H.A., Chan, L.H.M., Mahaffey, K.R., Murray, M., Sakamoto, M., Stern, H.A. 2007. Methylmercury exposure and health effects in humans: a worldwide concern. Ambio 36: 3-11.

39. ten Tusscher, G.W. 2002. Later childhood effects of perinatal exposure to background levels of dioxins in the Netherlands. PhD thesis. Amsterdam: University of Amsterdam.

40. J. A. Van Zorge, "Exposure to mixtures of chemical substances: is there a need for regulations?"Food and Chemical Toxicology, vol. 34, no. 11-12, pp. 1033–1036, 1996.

41. Filser, J., Koehler, H., Ruf, A., Rombke, J., Prinzing, A., Schaefer, M. 2008. Ecological theory meets soil ecotoxicology: challenge and chance. Basic and Applied Ecology 9: 346-355.

42. Van Gestel, C.A.M. 2008. Physico-chemical; and biological parameters determine metal bioavailability in soils. Science of the Total Environment 406: 385-395.

43. Sokolik, G.A., Ovsiannikova, S.V., Ivanova, T.G., Leinova, S.L. 2004. Soil-plant transfer of plutonium and americium in contaminated regions of Belarus after the Chernobyl catastrophe. Environment International 30: 939-947.

44. Zhao, C., Ren, J., Xue, C., Lin, E., 2005. Study on the relationship between soil selenium and plant selenium uptake. Plant and Soil 277: 197-206.

45. Liste, H., Prutz, I. 2006. Plant performance, dioxygenase-expressing rhizosphere bacteria, and biodegradation of weathered hydrocarbons in contaminated soil. Chemosphere 62: 1411-1420.

46. van der Geest, H, Leon Paumen, M. 2008. Dynamics of metal availability and toxicity in historically polluted floodplain sediments. Science of the Total Environment 406: 419-425.

47. Brouwer, A., Ahlborg, U.G., van den Berg, M., Birnbaum, L.S. 1995. Functional aspects of developmental toxicity of polyhalogenated hydrocarbons in experimental animals and human infants. European Journal of Pharmacology 293: 1-40.

48. Rutgers, M., Tuinstra, J., Spijker, J., Mesman, M., Wintersen, A., Posthuma, L. 2008. Risico's voor het ecosysteem in stap twee van het saneringscriterium [Ecosystem risks in step 2 of the remediation criterion]. Bilthoven (the Netherlands): RIVM.

49. Li, J., Zhou, B., Liu, Y., Yang, Q., Cai, W. 2008. Influence of coexisting contaminants on bisphenol A sorption and desorption in soil. Journal of Hazardous Materials 151: 389-393.

50. Gennings, C., Carter Jr, W.H., Casey, M., Moser, V., Carchman, R., Simmons, J.E. 2004. Analysis of functional effects of five pesticides using a ray design. Environmental Toxicology and Pharmacology 14: 115-126.

51. Hayes, T.B, Case, P., Chui, S., Chung, D., Haeffele, C., Haston, K., Lee, M., Mai, V.P., Marjuoa, Y., Parker, J., Tsui, M. 2006. Pesticide mixtures, endocrine disruption and amphibian declines: are we underestimating the impact. Environmental Health Perspectives 114, Supplement 1: 40-50.

52. Perry, M.J., Venners, S.A., Barr, D.B., Xu, X. 2007. Environmental pyrethoid and organophosphorous insecticide exposures and sperm concentration. Reproductive Toxicology 23: 113-118.

53. Laetz, C.A., Baldwin, D.H., Clllier, T.K., Hebert, V., Stark, J.D., Scholz, N.L. 2009. The synergistic toxicity of pesticide mixtures: implications for risk assessment and the conservation of the endangered Pacific salmon. Environmental Health Perspectives 117: 348-353.

54. Roos, P. H., Tschirbs, S., Pfeifer, F., et al., 2004. Risk potentials for humans of original and remediated PAH-contaminated soils: application of biomarkers of effect. Toxicology 205: 181–194.

55. Xiao, R. Y., Wang, Z., Wang, C. X., Yu, G., Zhu, Y. G. 2006. Genotoxic risk identification of soil contamination at a major industrialized city in northeast China by a combination of in vitro and in vivo bioassays. Environmental Science & Technology 40: 6170–6175.

56. Lee, D. H., Lim, J. S., Song, K., Boo, Y., Jacobs Jr., D. R. 2006. Graded associations of blood lead and urinary cadmium concentrations with oxidative-stress-related markers in the U.S. population: results from the third National Health and Nutrition Examination Survey. Environmental Health Perspectives 114: 350–354.

57. De Zwart, D. and Posthuma L. 2005. Complex mixture toxicity for single and multiple species: proposed methodologies. Environmental Toxicology and Chemistry 24: 2665-2676.

58. Altenburger, R., Walter, H., Grote, M. 2004. What contributes to the combined effect of a complex mixture? Environmental Science & Technology 38: 6353-6362.

59. Filser, J., Wittmann, R., Lang A. 2000. Response types in Collembola towards copper in the microenvironment. Environmental Pollution 107: 71-78.

60. Drexler, J.W. and Brattin, J.W. 2007. An in vitro procedure for estimation of lead relative bioavailability: with validation. Human and Ecological Risk Assessment 13: 383-401.

61. O'Halloran, K. 2006. Toxicological considerations of contaminants in the terrestrial environment for ecological risk assessment. Human and Ecological Risk Assessment 12: 74-83.

62. Römbke, J. 2006. Tools and techniques for the assessment of ecotoxicological impacts of contaminants in the terrestrial environment. Human and Ecological Risk Assessment 12: 84-101.

63. Roos, P.H., Tschirbs, S., Pfeifer, F., Welge, P., Hack, A., Wilhelm, M., Bolt, H.M. 2004. Risk potentials for humans of original and remediated PAH- contaminated soils: application of biomarkers of effect. Toxicology 205: 181-194.

64. Xiao, R., Wang, Z., Wang, C., Yu, G., Zhu, Y. 2006. Genotoxic risk identification of soil contamination at a major industrialized city in north east China by a combination of in vitro and in vivo bioassays. Environmental Science & Technology 40. 6170-6175.

65. Lee, D., Lim, J., Song, K., Boo, Y., Jacobs, D.R. Jr. 2006. Graded associations of blood lead and urinary cadmium concentrations with oxidative- stress-related markers in the US population: Results of the Third National Health and Nutrition Examination Survey. Environmental Health Perspectives 114: 350-354.

66. Neumann, H.G. 1996. Toxic equivalence factors, problems and limitations. Food and Chemical Toxicology 34: 1045-1051.

67. Dobrev, I.D., Andersen, M.E., Yang, R.S.H. 2002. In silico toxicology: simulating interaction thresholds for human exposure to mixtures of trichloroethylene, tetrachloroethylene and 1,1,1 trichloroethane. Environmental Health Perspectives 110: 1031-1039.

68. Silva, E., Rajapakse, N., Kortenkamp, A. 2002. Something from 'nothing'-eight weak estrogenic chemicals combined at concentrations below NOECs produce significant mixture effects. Environmental Science & Technology 36: 1751-1756.

69. Richter, M. and Escher, B. 2005. Mixture toxicity of reactive chemicals by using two bacterial growth assays as indicators of protein and DNA damage. Environmental Science & Technology 39: 8753-8761.

70. Caeiro, S., Costa, M. H., Ramos, T. B. 2005. Assessing heavy metal contamination in Sado Estuary sediment: An index analysis approach. Ecological Indicators 5: 151–169.

71. Håkanson, L. 1980. An ecological risk index for aquatic pollution control: A sedimentological approach. Water Research 14: 975–1001.

72. Long, E. R., MacDonald, D. D. 1998. Perspective: Recommended uses of empirically derived, sediment quality guidelines for marine and estuarine ecosystems. Human and Ecological Risk Assessment 4: 1019–1039.

73. Pekey, H., Karakas D., Ayberk, S., et al., 2004. Ecological risk assessment using trace elements from surface sediments of zmit Bay (Northeastern Marmara Sea) Turkey. Marine Pollution Bulletin 48: 946–953.

74. Roach, A. C. 2005. Assessment of metals in sediments from Lake Macquarie, New South Wales, Australia, using normalization models and sediment quality guidelines

75. Sutherland, R. A. 2000. Bed sediment-associated trace metals in an urban stream, Oahu, Hawaii. Environmental Geology 39: 611–627

76. Bhattacharya, A., Routh, J., Jacks, G., et al., 2006. Environmental assessment of abandoned mine tailings in Adak, Västerbotten District (Northern Sweden). Applied Geochemistry 21: 1760–1780

77. Usero, J., González-Regalado, E., Gracia, I. 1996. Trace metals in the Bivalve Mollusc Chamelea Gallina from the Atlantic Coast of Southern Spain. Marine Pollution Bulletin 32: 305–310.

78. Cheng, J. L., Shi, Z., Zhu, Y. W. 2007. Assessment and mapping of environmental quality in agricultural soils of Zhejiang Province, China. Journal of Environmental Sciences, 19: 50–54

79. Reeves, W.R., Barhoumi, R., Burghardt, R.C., Lemke, S.J., Mayura, K., McDonald, T.J., Phillips, T.D., Donelly, K.C.2001. Evaluation of methods for predicting the toxicity of polyaromatic hydrocarbon mixtures. Environmental Science & Technology 35: 1630-1636.

80. El Masri, H., Mumtaz, M.M., Yushak, M.L. 2004. Application of physiologically-based pharmacokinetic modeling to investigate the toxicological interaction between chlorpyrifos and parathion in rat. Environmental Toxicology and Pharmacology 16: 57 71.

81. Kortenkamp, A. and Altenburger, R. 1999. Approaches to assessing combination effects of estrogenic environmental pollutants. Science of the Total Environment 333: 131-140.

82. Sumpter, J.P. and Johnson, A.C. 2005. Lessons from endocrine disruption and their application to other issues concerning trace organics in the aquatic environment. Environmental Science & Technology 39: 4321-4332.

83. Brian, J.V, Harris, C.A., Scholze, M., Backhaus, T., Booy, P., Lamoree, M., Pojana, G., Jonkers, N., Runnalis, T., Bonfä, A., Marcomini, A., Sumpter, J. 2005. Accurate prediction of the response of freshwater fish to a mixture of estrogenic chemical. Environmental Health Perspectives 113: 721-728.

84. Wölfe, D. 1997. Interactions between 2,3,7,8 TCDD and PCs as tumor promoters: limitations of TEFs. Teratogenesis, Carcinogenesis and Mutagenesis 17: 217-224.

85. Marchant, C.A. 1996. Prediction of rodent carcinogenicity using the DEREK system for 30 chemicals currently being tested by the National Toxicology Program. Environmental Health Perspectives 104: 1065-1073.

86. Verhaar, H.J.M., Morroni, J.R., Reardon, K.F., Hays, S,M., Gaver, D.P., Carpenter, R.L., Yang, R.S.H. 1997. A proposed approach to study the toxicology of complex mixtures of petroleum products. Environmental Health Perspectives 195 (Supplement 1): 179-195.

87. Stevenson, C.N., Macmanus-Spencer, L.A., Luckenbach, T., Luthy, R.C., Epel, D. 2006. New perspectives on perfluorochemical ecotoxicology: inhibition and induction of an efflux transporter in the marine mussel Mytilus californianus. Environmental Science & Technology 40: 5580-5585.

88. Bailer, A.J., Hughes, M.R., See, K., Noble, R., Schaefer, R. 2002. A pooled response strategy for combining multiple lines of evidence to quantitatively estimate impact. Human and Ecological Risk Assessment 8:1597-1612..

89. Chapman, P.M., McDonald, B.G., Lawrence, G.S. 2002. Weight-of-evidence issues and frameworks for sediment quality (and other) assessments. Human and Ecological Risk Assessment 8:1489-1515.

90. Rutgers, M., den Besten, P. 2005. Approaches to legislation in a global context.The Netherlands perspective – soil and sediment. In:Toxicity Testing Environmental. Thompson C, Wahhia K, Loibner AP (eds). Blackwell, CRC Press, Oxford, pp 269–289.

91. Alexander, M. 1995. How toxic are toxic chemicals in soil? Environmental Science and Technology. 29:2713–2717.

92. Alexander, M. 2000. Aging, bioavailability, and overestimation of risk from environmental pollutants. Environmental Science and Technology. 34:4259–4265.

93. Callahan, C. A., Menzie, C. A., Burmaster, D. E.,Wilborn, D. C., Ernst, T. 1991 On-site methods for assssesing chemical impact on the soil environment using earthworms: a case study at the Baird and McGuire superfund site, Holbrook, Massachusetts. Environmental Toxicology and Chemistry. 10:817–826.

94. Attrill, M. J., Depledge, M. H., 1997. Community and population indicators of ecosystem health: Targeting links between levels of biological organisation. Aquatic Toxicology 38:183–197.

95. Keddy, C., Greene, J. C., Bonnell, M. A. 1994. A review of whole organism bioassays for assessing the quality of soil, freshwater sediment and freshwater in Canada. Scientific series no. 198. Environment Canada, Ottawa, Ontario, Canada.

96. Van Gestel, C. A. M., Kruidenier, M., Berg, M. P. 2003. Suitability of wheat straw decomposition, cotton strip degradation and bait-lamina feeding tests to determine soil invertebrate activity. Biology and Fertility of Soils 37:115-123.

97. Knacker, T., Förster, B., Römbke, J., Frampton, G.K. 2003. Assessing the effects of plant protection products on organic matter breakdown in arable fields – litter decomposition test systems. Soil Biology and Biochemistry 35:1269-1287.

98. Jager, T., Fleuren, R.H.L.J., Hogendoorn, E.A., De Korte, G. 2003. Elucidating the routes of exposure for organic chemicals in the earthworm, Eisenia andrei (Oligochaeta). Environmental Science and Technology 37:3399-3404.

Citations

CHAPTER 1

Helen K. French and Sjoerd E.A.T.M. van der Zee (2011). Research for Investigating and Managing Soil Contamination Caused by Winter Maintenance in Cold Regions, Soil Contamination, MSc Simone Pascucci (Ed.), ISBN: 978-953-307-647-8, InTech, DOI: 10.5772/23359.

CHAPTER 2

Eman M. ElSayed * and Shiv O. Prasher, Effect of the Presence of Nonionic Surfactant Brij35 on the Mobility of Metribuzin in Soil, doi:10.3390/app3020469

CHAPTER 3

Alejandro Rodriguez-Sanchez, Alejandro Gonzalez-Martinez , Maria Victoria Martinez-Toledo, Maria Jesus Garcia-Ruiz , Francisco Osorio and Jesus Gonzalez-Lopez, The Effect of Influent Characteristics and Operational Conditions over the Performance and Microbial Community Structure of Partial Nitritation Reactors, doi:10.3390/w6071905

CHAPTER 4

Sarah Roberts, Christopher Higgins and John McCray, Sorption of Emerging Organic Wastewater Contaminants to Four Soils, doi:10.3390/w6041028.

CHAPTER 5

Ivan Morales, Janet A. Atoyan, José A. Amador and Thomas Boving, Transport of Pathogen Surrogates in Soil Treatment Units: Numerical Modeling, doi:10.3390/w6040818#sthash.y5kl8AcD.dpuf.

CHAPTER 6

Charles Humphrey, Michael O'Driscoll and Jonathan Harris, Spatial Distribution of Fecal Indicator Bacteria in Groundwater beneath Two Large On-Site Wastewater Treatment Systems, doi:10.3390/w6030602;

CHAPTER 7

Yaqin Yu, Xiwu Lu and Yifeng Wu, Performance of an Anaerobic Baffled Filter Reactor in the Treatment of Algae-Laden Water and the Contribution of Granular Sludge, doi:10.3390/w6010122

CHAPTER 8

Ignacio Melendez-Pastor, M. Belén Almendro-Candel, Jose Navarro-Pedreño , Ignacio Gómez , Mariano García Lillo and Encarni I. Hernández, Monitoring Urban Wastewaters' Characteristics by Visible and Short Wave Near-Infrared Spectroscopy, doi:10.3390/w5042026

CHAPTER 9

Muhammad Aqeel Ashraf, Mohd. Jamil Maah and Ismail Yusoff (2014). Soil Contamination, Risk Assessment and Remediation, Environmental Risk Assessment of Soil Contamination, Dr. Maria C. Hernandez Soriano (Ed.), ISBN: 978-953-51-1235-8, InTech, DOI: 10.5772/57287.

Index

A

B

T

U

V

W